CARPENTRY
IN
COMMERCIAL
CONSTRUCTION

PRENTICE-HALL INTERNATIONAL, INC., London
PRENTICE-HALL OF AUSTRALIA PTY. LTD., Sydney
PRENTICE-HALL OF CANADA, LTD., Toronto
PRENTICE-HALL OF INDIA PRIVATE LIMITED, New Delhi
PRENTICE-HALL OF JAPAN, INC., Tokyo

PRENTICE-HALL, INC., Englewood Cliffs, New Jersey

Courtesy Symons Manufacturing Company

STANLEY BADZINSKI, JR.

Instructor
Milwaukee Area Technical College
Milwaukee, Wisconsin

CARPENTRY IN COMMERCIAL CONSTRUCTION

Library of Congress Cataloging in Publication Data

Badzinski, Stanley.
    Carpentry in commercial construction.

    Bibliography: p.
    1. Carpentry.   2. Concrete construction — Form work.
I. Title.
TH5604.B2   694   73-17446
ISBN 0-13-115212-2

© 1974 by
PRENTICE-HALL, INC.
Englewood Cliffs, New Jersey

All rights reserved. No part of this
book may be reproduced in any form
or by any means without permission
in writing from the publisher.

10 9 8 7 6 5 4 3 2 1

Printed in the United States of America

To the Student:

Carpentry, like all skilled trades, requires study and practice to attain proficiency. Do not minimize the importance of study if you wish to be a truly skilled carpenter.

As a young man contemplating a vocation, I was offered the following advice by my father, which I pass on to you: "... if you want to be a carpenter, study hard, and be a good one."

STANLEY BADZINSKI, JR.

CONTENTS

Preface ix

1. Carpentry In Commercial Construction 1
2. Forms for Concrete Footings 8
3. Forms for Concrete Walls 35
4. Forms for Concrete Columns 87
5. Forms for Concrete Beams and Girders 107
6. Forms for Concrete Floor Slabs 122
7. Forms for Concrete Stairways 155
8. Storefront Construction and Finishing 170
9. Movable Partitions 219
10. Cabinet and Fixture Work 230

Appendix A. Recommended Practice for Concrete Formwork 249

Appendix B. Requirements for Shoring Formwork 267

Appendix C. Manufacturers of Materials and Equipment 282

Selected Bibliography 286

Index 288

PREFACE

Carpentry in *Commercial Construction* was written for the carpenter apprentice engaged in concrete formwork and other phases of carpenter work on all types of commercial buildings and for students enrolled in architectural technology and construction technology courses at two-year technical institutes. It should also prove valuable to journeymen carpenters who wish to increase their knowledge of carpenter work on commercial buildings.

The introduction presents an overview of commercial carpentry for the new student of building construction. Materials used and the building of concrete forms for footings and concrete walls are discussed in detail. Various types of forms for reinforced concrete columns and beams are discussed and illustrated. Patented systems and conventional methods of building forms for reinforced concrete floor slabs are presented, with many illustrations and helpful hints on construction methods.

Concrete stairways offer some special forming problems. Examples and illustrations are included to make this phase of commercial carpentry easier to understand.

Storefront framing, which includes the building of display window areas, is included as is the installation of door bucks, conventional partitions, furring, plaster grounds, paneling, and acoustical ceilings.

Various types of movable partitions installed by carpenters in schools, office buildings, and other types of commercial buildings are illustrated, and typical installation details and procedures are discussed. Commercial cabinet and fixture work also is illustrated and discussed to complete the book's coverage.

I wish to take this opportunity to thank the various manufacturers, associations, and institutes for contributing information and illustrations that are used in this text. The illustrations used represent typical systems and in no way present all the similar products available. Therefore, whether products are included or excluded does not in any way reflect on their desirability.

I also wish to thank Mr. Stephen E. Cline for his encouragement in making this text a reality. Special thanks are due my wife, Alice, for the many hours she spent typing and proofreading the manuscript.

CARPENTRY IN COMMERCIAL CONSTRUCTION

CHAPTER ONE

Commercial construction as discussed in this book refers to all kinds of buildings other than homes and residential garages. It refers to retail stores, shopping centers, warehouses, factories, schools, and office buildings. This discussion of commercial construction will not consider the various kinds of buildings, but instead will be concerned with work done by carpenters on commercial buildings.

The carpenter is called on to perform many different types of work in the construction and interior finishing of a commercial building. The nature and amount of work performed by the carpenter will vary with the type and design of the building structure.

The carpenter's work on all buildings must be completed according to the plans and specifications provided for the job. Generally the carpenter follows details on the plans as work progresses, but the special instructions provided in the specifications cannot be overlooked. Usually the carpenter foreman and job superintendent are given the responsibility of making sure that the specifications are followed.

Building codes regulate the manner in which various structures can be built. These codes vary among different localities, and the carpenter must be familiar with the code in the locality in which he is performing his work. Knowledge of the code helps the carpenter avoid mistakes that could arise from incomplete plans, unclear specifications, or possible misinterpretation of the plans.

CONCRETE FORMING

Carpenters do a great deal of work on buildings constructed of reinforced concrete. Part of this work involves building forms in which the concrete is cast (see Fig. 1-1). These forms may be built on the job site or they may be prefabricated at a nearby carpenter shop. In some cases the forms are manufactured modular panels that are delivered to the job ready for erection and alignment.

FIG. 1-1. Forms for Concrete Construction.

Formwork can usually be divided into categories according to the phase of form construction, such as form building, form erecting, stripping, repairing, and dismantling.

Form building involves cutting stock materials such as boards, plywood, and framing lumber and assembling them into panels that are easy to transport and assemble. This work may be done in a shop away from the job site or in a temporary shop set up at the job.

Form erecting involves assembling and aligning prefabricated form panels into the size and shape required by the plans. Job-built forms are erected under the supervision of the carpenter foreman, the job superintendent, and sometimes the resident engineer. The amount of supervision required depends on the type of formwork and how complicated the shape of the form is. Patented forming systems are assembled according to the manufacturers' instructions under the carpenter foreman's supervision.

Stripping concrete forms involves removing forms from the cured concrete and cleaning them either for reuse on another part of the building or for storage until they are needed again.

After forms are stripped from the concrete, they are repaired as needed. All broken parts are replaced, and the forms are returned to service or storage as demanded by job requirements.

After the entire job is completed, the forms are dismantled. Patented form systems are cleaned and returned to storage. Job-built forms are taken apart, and all usable lumber is salvaged. All material that is damaged beyond repair is scrapped and disposed of in accordance with local ordinances.

Concrete Forming Systems

Concrete formwork may be divided into categories according to the type or system of formwork being used. Job-built forms are made of framing lumber, plywood, and form hardware.

Patented form systems made of steel frames with plywood sheathing are manufactured by several companies. These forms have their own special hardware for connecting panels, for tying, and for bracing.

Concrete forms made entirely of steel are also available. These, too, are patented systems. They are made in several flat and curved shapes and come with their own special type of connecting hardware. This type of form when properly used will give good service for many reuses.

Concrete forming for ribbed slab and waffle slab construction is usually done with steel or fiberglass reinforced plastic pans. These pans are available in a number of sizes and are installed on shoring especially prepared for them.

Special forms for stairs may be built from a combination of plywood, steel, and light framing lumber. Form liners, which are used to impart a special surface texture or design, are considered to be special forms and may be classified as a part of formwork for architectural concrete.

STORE FRONTS AND PARTITIONS

Building the display areas of storefronts is part of the work performed by carpenters on commercial buildings. This work involves building platforms and installing wood window frames, door frames, and doors. Because most storefronts are custom designed, highly skilled craftsmen are employed to perform the work (see Fig. 1-2).

FIG. 1-2. A Completed Storefront.

Some of the other work done by carpenters on commercial jobs includes building soffits above display windows, building conventional wood frame partitions, installing door bucks and arch centers for masonry walls, installing paneling, and installing acoustical ceilings.

MOVABLE PARTITIONS

Movable partitions made from various materials are used in many commercial buildings. These partitions make it possible to rearrange the work areas as the needs of the building's occupants change. Partitions made of a gypsum core with a vinyl covering are one type installed by carpenters (see Fig. 1-3).

FIG. 1-3. A Movable Partition. Courtesy Gold Bond Building Products, Division of National Gypsum Co.

A partition system made of enameled steel panels with a variety of connecting and installation hardware is popular in areas in which a hard surface and ease of cleaning are important. Other systems utilize steel hardware and a variety of wall surface materials.

CABINETS AND FIXTURES

Carpenters install all types of cabinets and fixtures in many types of commercial work, from stainless steel cabinets in hospitals and laboratories to cabinets made of wood and other materials in such structures as schools, factories, and retail stores (see Fig. 1-4).

FIG. 1-4. Typical Cabinet Installation.

Display cases and the wide variety of perimeter work are also installed by carpenters on retail store jobs. Perimeter work is decorative trim and shelving installed around the perimeter of the store. These items vary considerably in construction, and they are usually prefinished. Such features require craftsmen who are highly skilled at reading detail drawings and equally adept at handling and fitting prefinished material.

REVIEW QUESTIONS

1. What is commercial construction?
2. Why is the ability to read plans important in commercial construction?
3. How do building codes affect the carpenter's work?
4. List several types of work done by carpenters on commercial construction.
5. What are concrete forms?
6. What are the various categories or phases of formwork? Explain each.
7. What kinds of work are done by carpenters on storefronts?
8. What are some of the types of movable partitions installed by carpenters?
9. List the various kinds of fixture work done by carpenters.
10. What is perimeter work?

FORMS FOR CONCRETE FOOTINGS
CHAPTER TWO

Forms are required for wall footings and for column footings to give them their proper size and shape. These forms must be accurately located to provide a proper foundation for the walls and columns that their footings will support.

FORMS FOR COLUMN FOOTINGS

Column footing forms may take the shape of a rectangle, a square, or a truncated pyramid (see

FIG. 2-1. Column Footing Forms.

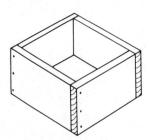
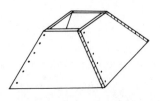
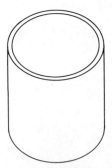

Typical form for square or rectangular column footing　　Pyramidal column footing form　　Circular column footing form

Fig. 2-1). They may be built from plywood, board lumber, framing lumber, fiber cylinders, and a variety of hardware and fasteners in the combination required to meet job needs.

Rectangular and Square Footings

Column footing forms in square and rectangular shapes may be built up from nominal 1 inch or 2 inch thick lumber. The side panels may be fabricated in such a way as to make the panels easy to strip from the cast concrete and reset for reuse.

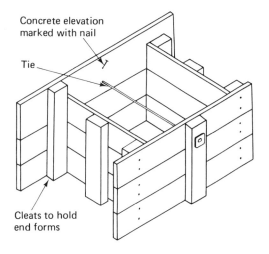

A. Box form with internal ties

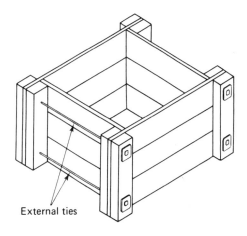

FIG. 2-2. Box Form for Column Footing.

B. Box form with external ties

The end panels of the form illustrated in Figure 2-2A are cut to the exact width of the footing. When placed in the form, these end panels serve as spacers for the side panels. A cleat is used to fasten the individual boards or planks together. In some cases these end panels may be made of plywood, thus eliminating the need for cleats and reducing the amount of work involved in building the form.

Side panels for this type of form are built up in a manner similar to the end panels, but the cleats function as a stop to keep the end panels from spreading. The distance between the cleats must be carefully maintained and should be equal to the footing size plus twice the sheathing thickness. The form panels can be left long because the location of the cleats regulates the size of the footing.

Snap ties or wire ties may be used as needed. Small forms usually will not require ties whereas larger forms require ties because of the heavier pressures of wet concrete. If ties are not used the forms may be held together by nails at the corners and by installing sufficient stakes to hold the form in place.

If many column footings of the same size must be formed, it is advantageous to build the forms with external ties (see Fig. 2-2B). Every part of a form built in this manner can be reused, and stripping the form from the hardened concrete is easy.

End panels for this form are cut to the exact size of the footing. Side panels are made longer to accommodate the cleats, which hold the side panels together and act as a backstop for the end panels. The distance between the cleats is determined by adding the combined thickness of the end form panels to the footing size.

The cleats are spaced properly and fastened to the form sheathing with nails that should be long enough to hold the cleats but not so long that they penetrate beyond the cleats. If nails penetrate the cleats, they should be clinched over to avoid accidents.

Ties used for this type of form can be any of a variety commercially available. All necessary hardware must be purchased to make the system work efficiently. Some of the types of ties available are discussed in Chapter 3 under *"Hardware."*

Circular Column Footings

Circular column footings may be formed with manufactured patented steel forms or from fiber tube forms cut to length at the job site (see Fig. 2-3). Steel forms are sturdy and can be reused many times when given reasonable care. Fiber forms are comparatively inexpensive, but they can be used only once. However, if the form can be permanently left in place, they may be more economical to use than some other types.

FIG. 2-3. Circular Column Footing Forms. Steel column forms are sometimes used to form footings. Fiber forms are cut to length on the job site.

Both steel and fiber forms are held in place with stakes, and the top of the form is usually set to grade (proper height in relationship to an established reference), thereby avoiding the need to establish a separate grade marking on each form. Setting the forms to grade also simplifies the placing of concrete, because the forms are simply filled to capacity.

Battered Column Footings

Forms for battered column footings are made from nominal 1 inch and 2 inch thick framing lumber, or from plywood and framing lumber. End panels for battered footings are made to the exact size of the footing, and the side panels are made larger to accommodate the cleats that support the end panels (see Fig. 2-4).

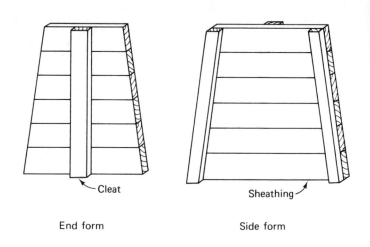

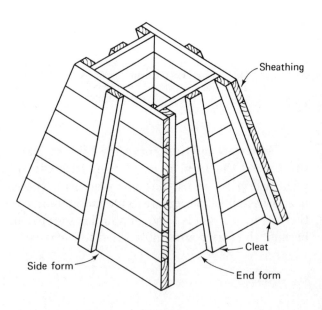

FIG. 2-4. Battered Footing Form.

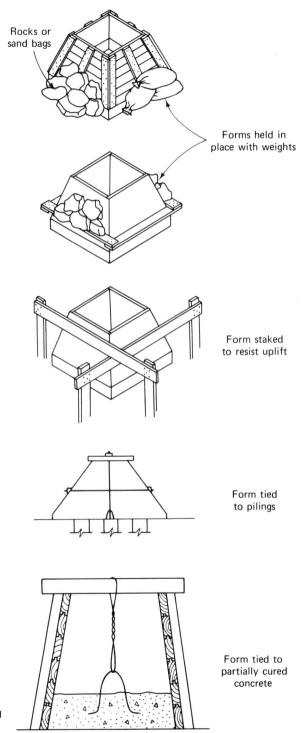

FIG. 2-5. Anchoring Battered Forms.

Concrete placed in forms of this shape exerts an upward pressure on the form. Therefore, to keep it from floating on the concrete, the form must be securely anchored in place by using stakes and braces; weights such as soil, sandbags, or rocks: or by tying the form to pilings or to anchors set in a partially cured lower portion of the footing (see Fig. 2-5).

Stepped Column Footing Forms

Large column footings that are very thick are often stepped to save on materials. Forms for this type of footing are made in the same manner as for square footings. That is, box forms of the proper size are built and set in place, with the smaller form set over the larger one (see Fig. 2-6). Various methods can be used to hold the upper form in place. One of the most common involves fastening a 2 by 4 to the lower edge of the top form and supporting it on the lower form (see Fig. 2-6).

If top forms are needed to keep the concrete from overflowing in the space between the upper and lower forms, the form must be anchored to resist the upward pressure of the wet concrete in the same manner as battered forms are secured in place.

FORMS FOR WALL FOOTINGS

Concrete forms for wall footings are usually built from 2 inch thick framing lumber of sufficient width. However, if the footings are exceptionally thick, the forms may be built using a combination of plywood, board sheathing, and 2 inch thick framing lumber; or manufactured form panels may be used to build the footing form in a manner similar to the way a wall form is built.

Plain Wall Footing Forms

The top of wall footing forms are set to grade in accordance with a grade stake or benchmark that has been established at the floor of the excavation. Inside edges of the form are placed at the footing line, and the distance between

FIG. 2-6. Stepped Column Footing Form.

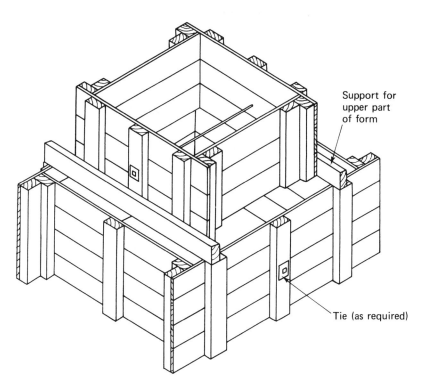

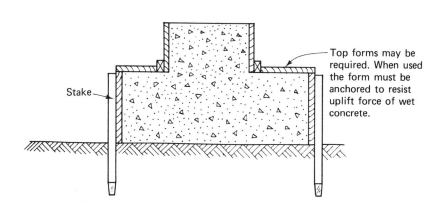

FORMS FOR CONCRETE FOOTINGS   16

the inside and outside forms is held to the footing width shown on the building plans.

Forms are held in place with stakes and braces (see Fig. 2-7A) that are driven at intervals of 4 to 8 feet. The spacing

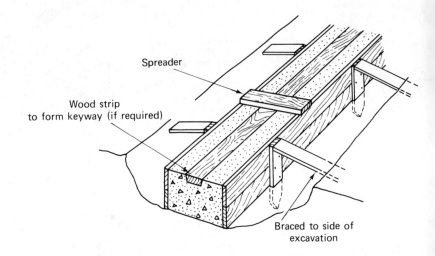

A. Form held in place with typical staking and bracing

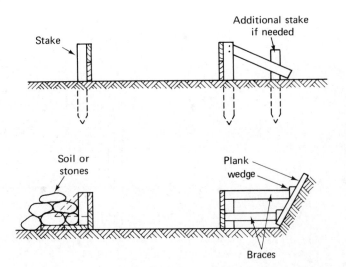

FIG. 2-7. Typical Wall Footing Form.

B. Extra staking or weights may be required by unusual soil conditions

between the stakes is governed by the thickness of the form material, the height of the form, and the type of soil in which the stakes are driven. The forms are nailed to the stakes with enough nails to keep them properly aligned while the concrete is being placed, but an excessive number of nails should not be used because this would make it difficult to remove the form. Usually one 8d or 12d nail per stake is sufficient, and it may not be necessary to nail the form to every stake.

If nominal 2 inch material such as 2 by 8 or 2 by 10, is used, fewer stakes are required to hold the form in place than if nominal 1 inch material were being used. The thicker material is more resistant to the pressure of fresh concrete and thus is less likely to bend.

As the height of the footing form increases, additional stakes and braces are needed to hold the form in alignment. If the form is near the side of the excavation and it is impossible to drive additional stakes, braces may be placed against a pad of 2 inch lumber bearing on the side of the excavation (see Fig. 2-7B).

Forms placed on soft or sandy soil require more stakes and braces than forms placed on medium soil because stakes are easily displaced. If forms are placed on very hard soil or rock and stakes cannot be driven, the forms may be held in place with braces and weights (see Fig. 2-7B).

Stepped Wall Footing Forms

If it is necessary to have the floor of the excavation at different elevations, a stepped footing form must be built (see Fig. 2-8). The number and height of the steps may be governed by either the plans and specifications or by job conditions or both.

The form for stepped footings is built in a manner similar to that for regular wall footings. In the stepped area the form is built so that the upper footings overlap the lower footings. Shutoffs are placed across the steps from one side form to the other. These shutoffs may be made from plywood, or from nominal 1 inch or 2 inch thick lumber. Two inch lumber is usually preferable because it will not require center bracing.

FORMS FOR CONCRETE FOOTINGS 18

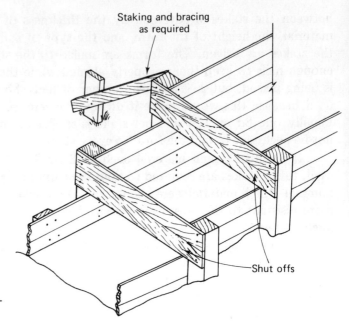

FIG. 2-8. Typical Stepped Footing Form.

FORMS FOR SLABS ON GRADE

Concrete slabs placed directly on grade (on the soil) for garage or dwelling floors, or for floors of large buildings, require side forms to outline their shape and location. These forms are similar in appearance to a one-sided wall footing form and are built in the same manner as wall footing forms.

LOCATING FOOTINGS

Footings and building lines are located on the floor of the excavation in accordance with previously established reference points. These reference points are often placed on batterboards.

Batterboards

Batterboards are L-shaped structures placed at the corners of the building, but far enough from the excavation so that they will not be disturbed during the excavating (see Fig. 2-9). To

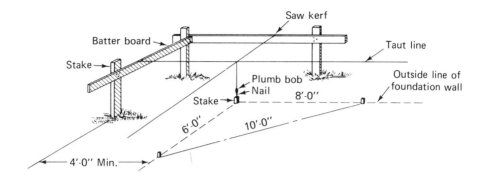

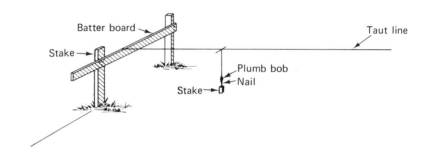

FIG. 2-9. Batterboards.

locate column lines, simple bar batterboards are used. These batterboards should be well braced so that building lines established on them will remain constant.

After the building lines are established, they can be permanently marked by cutting small saw kerfs in the batterboards. Nails can also be used to mark the building lines on the batterboards. Either of these methods is preferred to marking the location of the lines with a pencil because they are permanent and will not be washed away by rain or lost under dust and dirt.

Either a strong braided cotton or nylon line or a stranded wire can be stretched between the established marks on the batterboards. For rather long distances, usually over 100 feet, a stranded wire is usually more satisfactory because it can be stretched tighter, and therefore, will be less susceptible to movement by the wind.

FORMS FOR CONCRETE FOOTINGS 20

A plumb bob is used to locate points below the line (see Fig. 2-10). Care must be taken to keep the wind from moving the plumb bob away from its proper position. This can be done by shielding it and its line from the wind, by using a heavier plumb bob, or both.

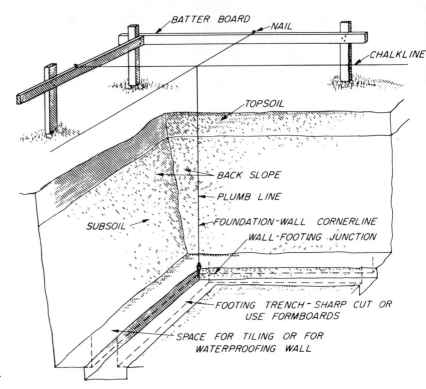

FIG. 2-10. Batterboards.

Grade Elevations

The heights of floors and footings are given as a grade elevation in feet and inches or feet and hundredths of a foot. The heights or grades are established from a known reference point, which is often referred to as a benchmark or datum point. The reference points may be established benchmarks based on sea level or on some stable object such as the intersection of concrete streets or sidewalks, a manhole cover, or a bolt on a fire hydrant. If grade elevations are given on a set of plans, the benchmark will also be stated.

When work is begun on the building project, additional benchmarks may be established for convenience. The grade elevation on these intermediate benchmarks is assigned in relation to the elevation of the original benchmark. A transit or builder's level and a surveyor's target rod or "story pole" are used to establish grade elevations. The transit can be used for reading grade elevations when set in a level position, or for reading angles of incline with the horizontal when in the transit position. Because it can be used in either the level or transit position, the transit is more commonly used than the builder's level, which can be used only for leveling or checking grade elevations.

THE LEVEL-TRANSIT

The level-transit is a precision instrument that is sturdily built but requires caution and care in handling. It is, therefore, necessary to avoid dropping the unit, bumping it against objects, or handling it in any careless manner whatsoever.

Level-transits made by different manufacturers have different features, but all should be kept clean, dry, and properly adjusted. When not in use they should be stored in the carrying case in accordance with the manufacturer's instructions. If the transit should need adjustment because of careless handling or an accident, it is best to return the unit to a qualified repair station where it can be repaired under controlled conditions.

Transit Nomenclature

Before attempting to use a transit, the prospective user should become familiar with the features of the instrument and its operation. Each different make or type of transit should be studied carefully before it is used.

The level-transit shown in Figs. 2-11 and 2-12 is one type commonly used on construction sites. If you study the illustration, you will be able to familiarize yourself with features common to most transits.

Telescope — The optical instrument through which sighting and reading is done. The magnifying power of the telescope in Fig. 2-11 is 26 times.

Objective Lens — The lens at the large end of the telescope.

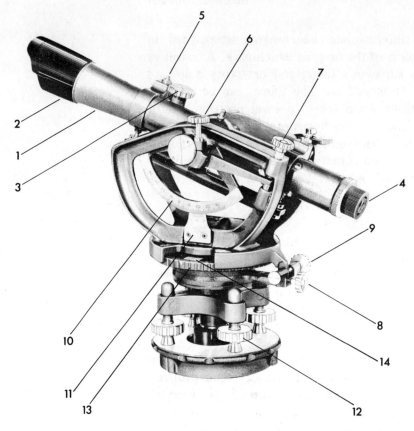

1. Telescope
2. Objective Lens
3. Focusing Knob
4. Eyepiece Cap
5. Crosshair Adjusting Screw
6. Vertical Motion Clamp
7. Vertical Motion Tangent Screw
8. Horizontal Motion Clamp
9. Horizontal Motion Tangent Screw
10. Vertical Arc
11. Vertical Arc Vernier
12. Leveling Screws
13. Horizontal Circle
14. Horizontal Circle Vernier

FIG. 2-11. Transit Nomenclature. Courtesy of David White Instruments, Division of Realist, Inc.

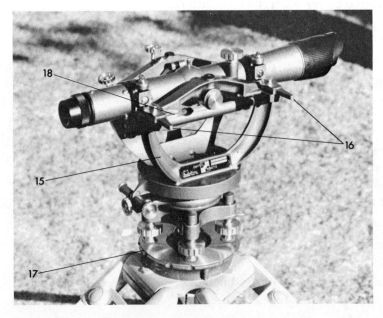

FIG. 2-12. Transit Nomenclature (continued).

15. Standards
16. Telescope Locking Levers
17. Leveling Baseplate
18. Telescope Level Tube

Focusing Knob — The knob on top of the telescope used to focus the scope on objects at different distances.

Eyepiece Cap — The Cap over the eyepiece that is rotated to focus the crosshairs.

Crosshair Adjusting Screw — The screws used to adjust the crosshairs. Crosshairs should be adjusted at qualified repair stations.

Standards — The frame that supports the telescope.

Telescope Locking Levers — The levers used to lock the telescope at right angles to the vertical axis of the instrument. The levers are unlocked to use the instrument in transit position.

Horizontal Motion Clamp — The screw used to lock horizontal motion of the transit.

Horizontal Motion Tangent Screw — The screw used to make fine horizontal adjustments after the horizontal motion is locked.

Horizontal Circle — The circle at the base of the instrument that is graduated in degrees.

Horizontal Circle Vernier — The scale used for reading fractions of a degree.

Leveling Baseplate — The plate that attaches the transit to the tripod and serves as a base for the leveling screws.

Leveling Screws — The screws that are adjusted to level the instrument.

Vertical Motion Clamp — The screw used to lock vertical motion of the telescope.

Vertical Motion Tangent Screw — The screw used to make fine vertical adjustments after the vertical motion clamp is locked.

Vertical Arc — The arc graduated in degrees for measuring upward and downward angles.

Vertical Arc Vernier — The vernier used to read parts of a degree on the vertical arc.

Telescope Level Tube — The level tube attached to the telescope that is used to level the instrument.

Tripod — The three-legged support on which the transit is installed for use.

FORMS FOR CONCRETE FOOTINGS 24

Transit Setup

The tripod should be set up with the legs set firmly before the transit is removed from the carrying case. As a general rule, the legs should have a spread of about $3\frac{1}{2}$ feet (see Fig. 2-13), and the head of the tripod should appear to be level. The legs of the tripod should be fastened securely to the tripod head, and the adjustable legs should be tightened to avoid slippage when the transit is placed on the tripod.

FIG. 2-13. Transit on Tripod.

Before removing the instrument from the case, the horizontal motion clamp should be loosened. Then, grasping the instrument firmly in one hand, the leveling baseplate can be turned free of the carrying case.

Carefully place the instrument on the tripod and fasten it securely by turning the leveling baseplate, but do not overtighten it. If the instrument is to be used to measure an angle, the plumb bob must be attached to the hook or chain provided at the base of the transit and adjusted until it is just above grade. If it is necessary to move the instrument to bring it directly over the point, this must be done before leveling the transit. A small amount of movement can be obtained by loosening two adjacent leveling screws and moving the transit on the leveling baseplate. If the transit still is not over the point, the tripod must be moved.

When the tripod is finally positioned, be sure that each of the points are well into the ground so that they will not move. On paved surfaces, in which points cannot be driven, be sure that they are positioned to hold securely.

Leveling the Instrument. The most important operation in preparing the instrument for use is leveling. A poorly leveled instrument will give false readings, and the amount of error will increase as the distance from the transit to the target increases.

The first step in leveling is to be sure that the instrument is locked in place on the baseplate by checking the tightness of two adjacent leveling screws. When leveling the transit, two opposite leveling screws must be turned the same amount at the same time. The direction in which they are turned is such that the thumb on each screw moves in at the same time or moves out at the same time (see Fig. 2-14).

FIG. 2-14. Leveling the Transit.

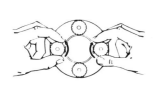
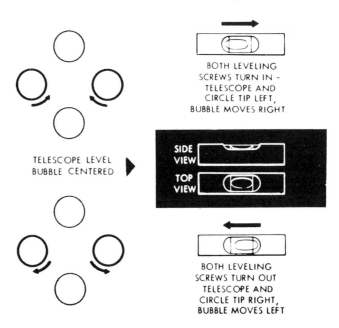

After checking to be sure the instrument is locked in place, the actual leveling is begun by turning the telescope so that it is lined up across an opposite pair of leveling screws. The bubble in the level vial is brought to the center of the vial by adjusting the leveling screws over which the telescope was set. The bubble moves in the direction of the left thumb. Therefore, if the bubble must be moved to the right, the left thumb must move in to the right; but if the bubble must move to the left, the left thumb must move out to the left.

When the bubble is centered in the vial with the telescope over one opposite pair of leveling screws, the telescope is rotated 90° so that it is over the other pair of leveling screws. This pair of leveling screws is adjusted to bring the bubble to the center of the vial. Then the telescope is turned across the first pair of screws and readjusted. After rechecking the level across both pairs of screws, it should be possible to rotate the telescope 360° without changing the position of the bubble.

If the level vial is out of adjustment, it is still possible to level the instrument by leveling the telescope over one pair of leveling screws and then rotating the telescope 180°. The leveling screws are adjusted to bring the bubble back one-half the distance it moved off center. Then the telescope is rotated 90° and the bubble is brought to the same relative position it occupied before the telescope was turned. After rechecking, it should be possible to rotate the telescope in a complete circle without changing the position of the bubble in the level vial.

Measuring Differences in Grade Elevation

The transit can be used to measure differences in grade elevations and to transfer grade elevations with the aid of a target rod. In using the transit for this purpose, the tripod must be well set and the transit must be leveled accurately. The target rod must be held perfectly vertical. Side-to-side movement can be checked along the vertical crosshair. The smallest reading is obtained when the rod is held vertically. Therefore, if the rod is moved back and forth, the smallest reading is the one recorded.

The first reading taken at the benchmark is called backsight (BS+), and when added to the elevation at the benchmark (BM) it establishes the elevation of the line of sight and is called the height of the instrument (HI).

Readings taken to other points after the height of the instrument is established are called foresight (FS-). To find the elevation at these other points, the foresight is subtracted from the height of the instrument previously established (see Fig. 2-15). This newly established elevation is marked on the new stake or benchmark and on the proper line of the elevation record notebook. A sample page of a typical record is shown in Fig. 2-16.

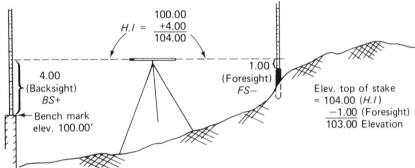

FIG. 2-15. Finding Backsight and Foresight.

Station	BS+	H1	FS−	Elev
BM				100.00
1	3.75	103.75		
A		103.75	12.50	91.25
2	5.25	96.50		
B		96.50	4.50	92.00

FIG. 2-16. Record of Elevations.

Transferring Grade Elevations

One of the most common uses of the level transit on commercial jobs is for establishing footing elevations at the floor of the excavation and for checking the elevations as the forms are placed. The procedure for transferring grade elevations is explained in the following example.

FORMS FOR CONCRETE FOOTINGS

To establish a benchmark at the floor of an excavation, the transit is set up at some convenient point within 100 feet of the benchmark and the excavation (see Fig. 2-17). The backsight reading from station 1 to the benchmark is 3.75 feet. Therefore, the height of the instrument is 103.75 feet. The instrument is turned and a foresight reading taken at temporary stake A is found to be 12.50 feet. The grade at the top of stake A is established by subtracting 12.50 feet (FS) from 103.75 feet (HI), or 91.25 feet.

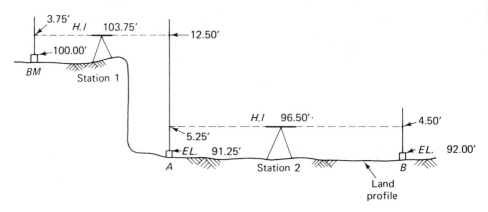

FIG. 2-17. Transferring Grade Elevations.

The instrument must now be moved to station 2 at the floor of the excavation (see Fig. 2-17) and be set up and leveled before elevations of other grade stakes can be established. The backsight taken from station 2 to stake A establishes the new height of the instrument (HI) at 96.50 feet. A foresight (FS-) reading taken to stake B is 4.50 feet and when subtracted from 96.5 feet (HI), the elevation of stake B is established at 92.00 feet.

Layout of Building Lines

To use the transit to lay out building lines, one corner of the building must be located with a stake, and a line representing one side of the building must be established. In Fig. 2-18 the transit is set up at stake A, and with the aid of a plumb bob is centered over the nail in the stake that represents the exact corner. The transit is then leveled in the usual manner.

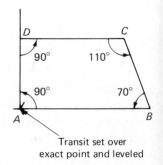

FIG. 2-18. Turning an Angle.

After leveling, the telescope locking levers are opened and a sighting is taken on the establsihed line near point D. To help in sighting point D, the horizontal motion may be locked and the horizontal tangent screw adjusted to sight point D accurately.

With the horizontal motion locked and the vertical crosshair set on point D, the horizontal circle may be turned to 0°. Now the horizontal motion may be unlocked and the transit turned 90° toward point B. After turning the instrument, the horizontal motion is again locked. With the telescope in transit position, stake B is located at the proper distance from stake A. A nail is driven in the top of stake B to locate the exact corner.

After checking the layout, the transit is moved to point B and located and leveled in the same manner as at point A. After leveling, a sighting is taken on point A. The horizontal motion is locked, and the horizontal circle is set to 0°. Then the horizontal motion is unclamped, and the transit is rotated 70°. Stake C is then located in the same manner as was stake B.

To continue the layout, the transit is moved to stake C and leveled in the usual manner. The vertical crosshair of the telescope is aligned on stake B, and the horizontal motion is again clamped. After the horizontal circle is set to 0°, the horizontal motion is unclamped and the transit is rotated 110°. Sighting through the transit, the vertical crosshair should fall on stake D.

Should any serious errors exist after completing the layout, a check of dimensions should first be made and, if necessary, the entire layout should be repeated to reconcile the errors.

Reading the Vernier Scale

All level transits are equipped with a vernier scale that enables readings to be made in parts of a degree. The vernier scale and the obtainable accuracy will vary among different manufacturers, but the procedure in reading the scale is similar for all types.

The vernier in Fig. 2-19 divides each degree into 12 equal parts of 5 minutes each. Although the carpenter will seldom need to read angles in parts of a degree, he should know how to use the vernier scale. The vernier illustrated is marked 60 in the center, which is also 0. In reading the vernier, start at 0 and read up the scale until a mark on the vernier and the horizontal circle coincide. The illustration on the top reads 77° 20 minutes.

In reading up the scale, if no mark coincides when we reach 30 minutes on the right, move to the 30 on the left and continue reading the scale from left to right. The lower illustration in Fig. 2-19 reads 76° 45 minutes.

Establishing Points on a Line

The level-transit can be used to accurately establish points along a line between established reference points. When used for this purpose, the instrument must be set up directly over an established point and sighted in and locked on a second reference point. If it is not possible to set up over a reference point, the instrument must be set up so that it lies on a straight line extended from the established reference points. Locating the instrument along this line is a trial-and-error process, often referred to as "wiggle in."

To run a line with the instrument set over an established reference point, the instrument is first located exactly over the point on the top of the stake and then carefully leveled. With the instrument in transit position, it is sighted on the distant reference point, and with the horizontal motion screw locked, the horizontal tangent adjusting screw may be used to bring the vertical crosshair exactly on the reference point. After this is done, the telescope may be rotated in a vertical plane and focused at the various points where additional stakes are placed (see Fig. 2-20).

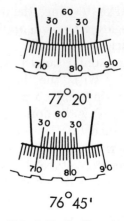

FIG. 2-19. Reading the Vernier Courtesy David White Instruments, Division of Realist, Inc.

FIG. 2-20. Running a Line. Courtesy David White Instruments, Division of Realist, Inc.

If it is necessary to wiggle in, skill and judgment on the part of the operator is important. The first step is to set up and level the instrument at a point judged to be in line with the reference points. Next, the instrument is sighted on the nearest reference and the horizontal motion is locked. Now it is turned in a vertical plane and sighted over the next reference point without changing the horizontal motion. Almost always the line of sight will not cross the reference point, but will miss it by some distance (see Fig. 2-21A).

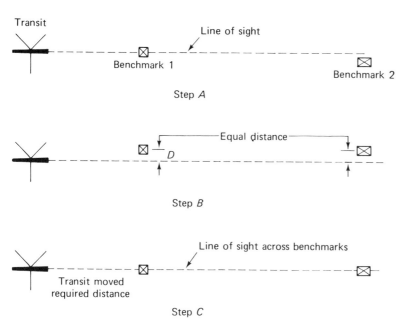

FIG. 2-21. "Wiggling In" — Plan View.

The next procedure is to adjust the horizontal motion so that the line of sight is parallel to the reference points (see Fig. 2-21B). This adjusting also is a trial-and-error procedure in which the end of a wood folding rule or steel tape is held on the reference points and the transit horizontal motion is adjusted until equal readings are taken at both references. This reading is the distance the transit must be moved to bring the line of sight in alignment with the established references.

After moving the transit the required distance, it is releveled and sighted on the nearest reference point. With the vertical crosshair on the reference point and the horizontal motion locked, the telescope is rotated in a vertical plane to check the alignment with the far reference point. If the vertical crosshair does not fall on the reference, the steps outlined in the preceding paragraph must be repeated.

When the alignment is correct, there will be a straight line of sight from the transit through the reference points, as shown in Fig. 2-21C. The transit may then be used to establish any number of additional points on the line in much the same manner as shown in Fig. 2-20.

STAKING OUT

After building lines and corner points have been established, it is often advantageous to lay out corners using the 3-4-5 rule. The numbers 3, 4, and 5 represent the lengths of the sides of a right triangle. Any multiple of 3, 4, and 5 may be used (see Table 2-1), and it is best to use the largest multiple possible to make the completed layout more accurate.

Table 2-1. Multiples of 3-4-5

a	b	c	a	b	c
3	4	5	18	24	30
6	8	10	21	28	35
9	12	15	24	32	40
12	16	20	27	36	45
15	20	25	30	40	50

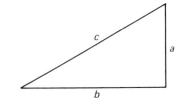

To use the 3-4-5 rule, a building line and the location of a corner must previously have been established. Starting from the established corner, the length of one side of the 3-4-5 triangle is marked off on the established line. The length of the other side is measured off at approximately 90° to the established line. The third side of the triangle is marked off from the established line to the length of the second side. Where the two dimensions come together, the third corner of the triangle is located. This procedure is illustrated and summarized in Fig. 2-22.

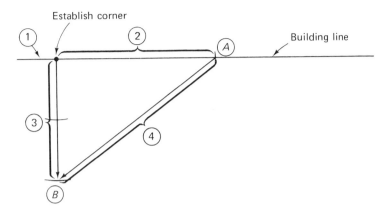

(1) Corner and building line must be established.
(2) Mark off length of one side of triangle.
(3) Mark off length of second side of triangle at approximately 90° to first side.
(4) Starting at "A" mark off length of the hypotenuse.

NOTE: Using two measuring tapes the second and third sides may be measured simultaneously. Corner B is established where dimensions 3 and 4 coincide.

FIG. 2-22. 3-4-5 Layout.

REVIEW QUESTIONS

1. What function do footing forms perform?
2. Sketch two types of column footing forms.
3. What materials may be used to build column footing forms?
4. How are circular column footings formed?
5. What is a battered footing?
6. What precautions must be observed when placing battered footing forms?
7. How are wall footings formed?
8. How are wall footing forms located?
9. How are wall footing forms held in place?
10. What are batterboards?
11. How are grade elevations determined?

12. What is a benchmark?
13. What is a level-transit?
14. Outline the procedure for setting up and leveling a transit.
15. How are differences in grade elevations measured?
16. What is backsight? foresight? height of instrument?
17. Outline the procedure for laying out building lines.
18. Outline the procedure for establishing points along a line with the transit.
19. Outline the procedure for wiggling in.
20. Outline the use of the 3-4-5 rule for laying out a square corner.

FORMS FOR CONCRETE WALLS
CHAPTER THREE

Concrete walls of all types require some type of formwork to hold the concrete while it is wet or plastic. These forms are actually molds that give the concrete its required shape, and sometimes they also impart the desired surface finish to the concrete.

Wall forms may be built on the job site or in a nearby carpenter shop from standard framing lumber, plywood, and an assortment of hardware. They may also be assembled from a variety of patented form panels manufactured by a number of different companies.

FORMWORK MATERIALS

A variety of standard construction materials are used in building concrete forms, including framing lumber (2 by 4's, 2 by 6's, etc.), common boards, plywood, nails, bolts, and wire. In addition, a variety of specialty hardware is used for ties, spreaders, clamps, and braces.

Lumber

Almost any framing and sheathing lumber that is available locally may be used for form construction. Occasionally, certain lumber grades or species will be required to meet strength or surface finish requirements. It is not the carpenter's responsibility to choose the grade of lumber for the forms. His responsibility is to build the forms in conformance with the plans and specifications.

Framing lumber may be surfaced green or surfaced dry. If the lumber is surfaced with a moisture content of more than 19 percent, it is considered green and is finished to a larger size. If it has a moisture content of 19 percent or less, it is considered dry and is surfaced to a smaller size. Standard sizes for lumber surfaced either green or dry are given in Table 3-1.

Table 3-1. Standard Lumber Sizes in Inches Based on CS 20-70

NOMINAL SIZE	SURFACED GREEN	SURFACED DRY
2 × 4	$1\frac{9}{16} \times 3\frac{9}{16}$	$1\frac{1}{2} \times 3\frac{1}{2}$
2 × 6	$1\frac{9}{16} \times 5\frac{5}{8}$	$1\frac{1}{2} \times 5\frac{1}{2}$
2 × 8	$1\frac{9}{16} \times 7\frac{1}{2}$	$1\frac{1}{2} \times 7\frac{1}{4}$
2 × 10	$1\frac{9}{16} \times 9\frac{1}{2}$	$1\frac{1}{2} \times 9\frac{1}{4}$
2 × 12	$1\frac{9}{16} \times 11\frac{1}{2}$	$1\frac{1}{2} \times 11\frac{1}{4}$
4 × 4	$3\frac{9}{16} \times 3\frac{9}{16}$	$3\frac{1}{2} \times 3\frac{1}{2}$
4 × 6	$3\frac{9}{16} \times 5\frac{5}{8}$	$3\frac{1}{2} \times 5\frac{1}{2}$
4 × 8	$3\frac{9}{16} \times 7\frac{1}{2}$	$3\frac{1}{2} \times 7\frac{1}{4}$

Board lumber is used for form sheathing, ledgers, bracing, and blocking. It may be purchased rough or surfaced, and may be seasoned or unseasoned. Lumber that is subjected to heavy moisture should not be seasoned. Lumber used for form sheathing will swell (expand) when subjected to wet concrete. Therefore, to avoid form distortion, unseasoned or partially seasoned lumber is preferable.

Surfaced boards may be purchased S4S (square edge), shiplapped, or tongue and groove. The surface area they cover varies with the width and type of edge treatment. Therefore, if board sheathing is used, the area to be covered must be multiplied by a factor that makes allowance for the material lost during surfacing. Knowledge of these factors is important when ordering material for form sheathing (see Table 3-2).

Table 3-2. Sheathing Coverage Estimator

SHIPLAP	Nominal Size	WIDTH		AREA FACTOR*	TONGUE AND GROOVE	Nominal Size	WIDTH		AREA FACTOR*	S4S	Nominal Size	WIDTH		AREA FACTOR*
		Dress	Face				Dress	Face				Dress	Face	
	1 x 6	5½	5⅛	1.17		1 x 4	3⅜	3⅛	1.28		1 x 4	3½	3½	1.14
	1 x 8	7¼	6⅞	1.16		1 x 6	5⅜	5⅛	1.17		1 x 6	5½	5½	1.09
	1 x 10	9¼	8⅞	1.13		1 x 8	7⅛	6⅞	1.16		1 x 8	7¼	7¼	1.10
	1 x 12	11¼	10⅞	1.10		1 x 10	9⅛	8⅞	1.13		1 x 10	9¼	9¼	1.08
						1 x 12	11⅛	10⅞	1.10		1 x 12	11¼	11¼	1.07

*Allowance for trim and waste should be added.

EXAMPLE

A wall form panel 10' high and 48' long would have an area of 480 sq ft. The board feet of lumber required for sheathing can be determined by multiplying the form area by an area factor from Table 3-2.

Using 1 × 6 shiplap boards:
480 sq ft × 1.17 = 562 board ft

Using 1 × 6 tongue and groove boards:
480 sq ft × 1.17 = 562 board ft

Using 1 × 6 S4S boards:
480 sq ft × 1.09 = 523 board ft

In each case an additional allowance of 2% to 10% should be made for waste in cutting and fitting.

Plywood

On many concrete forming jobs plywood has replaced solid board lumber as a form sheathing material. Plywood has the advantage of fewer joints and faster fabrication. Almost any grade of plywood manufactured for exterior use is suitable for concrete forming. However, the grading rules of the American Plywood Association recognize the special requirements of plywood used for concrete forming, and a number of special Plyform grades have been established.

All Plyform grades are manufactured with B and C or C plugged veneers and all are made with waterproof glue. Table 3-3 summarizes the characteristics and uses for Plyform. Class 1 Plyform is manufactured with group 1 veneers and is stronger and stiffer than class II Plyform, which may be made up of

Table 3-3. American Plywood Association Plyform Grade-use Guide

Grade-Use Guide for Concrete Forms*				
Use these symbols when you specify plywood	Description	Typical grade-trademarks	Veneer Grade	
			Faces	Inner plys
B-B Plyform Class I & Class II**	Specifically manufactured for concrete forms. Yields many reuses. Smooth, solid surfaces. Edge-sealed. Mill-oiled unless otherwise specified.	B-B PLYFORM CLASS I EXTERIOR DFPA	B	C
HDO Plyform Class I & Class II**	Hard, semi-opaque resin-fiber overlay, heat-fused to panel faces. Smooth surface resists abrasion. Yields up to 200 reuses. Edge-sealed. Light oiling recommended after each pour.	HDO·PLYFORM· EXT·DFPA·PS1·66	B	C Plugged
Structural I Plyform	Especially designed for engineered applications. Contains all Group I species. Stronger and stiffer than Plyform Class I and II. Especially recommended for high pressures where face grain is parallel to supports. Also available with HD Overlay.	STRUCTURAL I B-B PLYFORM CLASS I EXTERIOR DFPA	B	C or "C" Plugged
Special Overlays, Proprietary panels and MDO plywood specifically designed for concrete forming.**	Panels produce a smooth uniform concrete surface. Generally mill treated with form release agent. Check with manufacturer for design specifications, proper use, and surface treatment recommendations for greatest number of reuses.			
*Commonly available in $\frac{5}{8}$" and $\frac{3}{4}$" panel thicknesses (4' x 8' size).				
**Check dealer for availability in your area.				

1-66. All B-B Plyform is oiled on both faces and edge sealed at the mill. It may be special ordered with no face oil and edge sealed only.

The purpose of the edge sealing and face oiling is to prevent or at least reduce the amount of moisture entering the plywood when concrete is being placed. Face oiling also acts as a release agent when stripping the forms from the concrete. However, unless the mill oiling is reasonably fresh at the time of initial use, job oiling is required. For this reason, many contractors prefer to order plywood for concrete forming edge sealed only. They then can apply the necessary sealer or release compound at the appropriate time on the job site.

The high density overlay (HDO) on plywood is smooth and waterproof. It is not necessary to oil these panels for initial use, but to make form removal easier most users apply a light coating of oil or release agent.

Form Release Agents

Form release agents permit clean and easy release of the form from the hardened concrete. They must leave the surface of the concrete hard and durable, and they must provide protection for the form material to extend its useful life.

Form release agents can be classified as straight oil, emulsified waxes, oil emulsions, volatile coatings, chemically active coatings, and resinous coatings. Each has different characteristics that make them desirable for certain types of work.

A number of companies manufacture and market form release agents. Some of the commonly used release agents are Symons Magic Kote, Allenform Coating, Uni-Form Oil, and plain mineral oil. Magic Kote and Uni-Form release agents are sold in 55 gallon drums and are ready for use when delivered to the job. Allenform Coating is a concentrate sold in 1 quart cans. One quart mixed with 55 gallons of No. 2 diesel fuel makes an inexpensive form coating material.

Form release agents are usually applied by roller, brush, or sprayer. They should be applied uniformly to obtain consistently good results. The product manufacturer's recommendations for method and rate of application should be carefully followed, and in all cases puddles of excess oil should be wiped off the form before the forms are put in place.

Nails and Wire

A variety of nails are used in building concrete forms. Formwork that is permanently fastened together may be nailed with common nails, sinker nails, or box nails. Form parts that are fastened temporarily should be nailed together with double-headed nails (see Fig. 3-1).

Form sheathing may be fastened to the studs with common nails, box nails, or other types with a large head. The nail used should be long enough to penetrate at least 1 inch into the studding. Some contractors use 5d nails for $\frac{5}{8}$ inch plywood

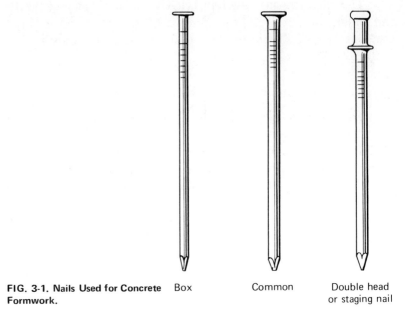

FIG. 3-1. Nails Used for Concrete Formwork.   Box   Common   Double head or staging nail

sheathing and 6d nails for $\frac{3}{4}$ inch plywood sheathing. If the studding does not hold nails well, a longer nail is used.

Annealed wire is used for tying forms and also for providing ties in other types of commercial carpentry. This wire is strong and fairly soft. It can be twisted around form members and ties by twisting the ends together. A number 12 SWG (Steel Wire Gauge) wire, which is 0.1055 inches in diameter, is most often used. In some lightweight operations a smaller number 16 SWG wire is used. This wire is 0.0625 inches in diameter and can be more easily bent and tied than the number 12 wire. Hardened or annealed wire and rods (wire supplied in coils, rods in straight lengths) $\frac{1}{4}$ inch, $\frac{3}{8}$ inch, and $\frac{1}{2}$ inch in diameter are also used in a variety of ways for form ties.

Hardware

Form hardware is available from a number of companies in a wide variety of types for various applications in wall form construction. Some of the more common types will be discussed in the following paragraphs. The reader is reminded to check the catalogs of various manufacturers and suppliers for additional information.

Snap ties are commonly used in form construction (see Fig. 3-2). They are made for a variety of wall thicknesses and

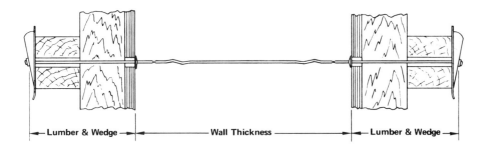

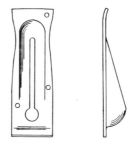

FIG. 3-2. Snap Tie and Wedges. Courtesy Dayton Sure-Grip & Shore Co.

Steel Wedges

form makeup. The snap tie wire is crimped to hold spreader washers at the proper spacing. Additional crimping is sometimes used to keep the tie from turning in the concrete during the breakback operation.

Most ties are manufactured so that they will break off inside the wall 1 inch from the finished surface, but other breakbacks are available. The breakback point on the tie is created by reducing the cross-sectional area of the tie. This reduced section may be placed $\frac{1}{4}$ inch to 1 inch from the wall surface. A special wax or paint coating is applied to the tie from the reduced section to the spreader washer to prevent the tie from bonding to the concrete. This bond breaker assures a positive breakback. Snap ties generally breakback easiest if they can be removed within 24 hours after the concrete is placed.

Wood or plastic cones may be used in place of spreader washers to assure proper breakback, proper form spacing, and to prevent grout (mixture of cement and water) leakage (see Fig. 3-3). This type of tie is used most often in architectural

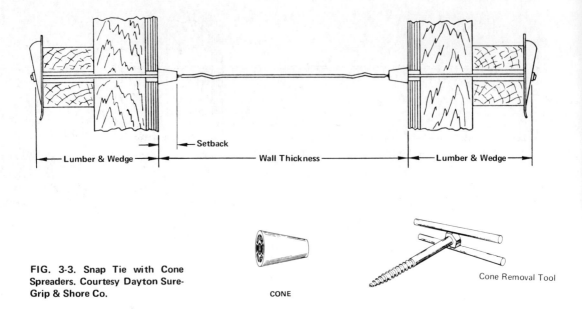

FIG. 3-3. Snap Tie with Cone Spreaders. Courtesy Dayton Sure-Grip & Shore Co.

concrete in which the holes created by the cone may either be filled to provide a smooth surface or may be left open to provide a decorative effect. These ties are available with a 1 inch, $1\frac{1}{2}$ inch, or 2 inch setback.

The cones should be removed from the concrete as soon as possible after the form is removed to avoid staining the concrete. Wood cones will be easier to remove if left exposed after the form has been removed and allowed to dry out for a short time. A cone removal tool is available from tie manufacturing companies to help in removing the cones (see Fig. 3-3).

On architectural concrete, it may be desirable to use stainless steel ties, which eliminate the possibility of rust stains from exposed ties. Stainless steel ties are generally available in the same sizes as plain steel ties and with any size cone spreader.

Coil ties are generally used on larger heavy duty forms. They are made with either two or four struts, depending on strength requirements, and are used with coil bolts, large flat washers, and spreader cones if required (see Fig. 3-4). Coil ties are versatile and economical in many applications, and they provide a positive disconnect system that makes stripping large crane handled forms easier than stripping some other tie systems.

B-1 TWO STRUT

B-2 FOUR STRUT

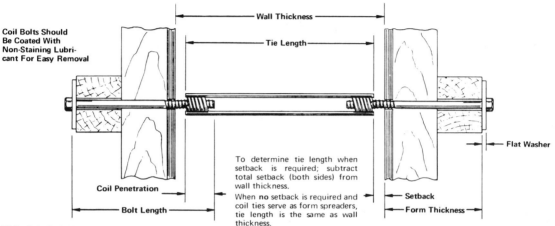

FIG. 3-4. Coil Ties. Courtesy Dayton Sure-Grip & Shore Co.

Various accessories are available for use with coil ties. These include various types of setback cones, coil bolts, coil threaded rods, coil thread nuts, flat washers, and batter washers, and are illustrated in Fig. 3-5. Other coil tie accessories are available, and the reader is advised to check various manufacturers' catalogs for other items.

She bolts are another tie system used on large heavy-duty forms. Standard she bolts are 20 inches long, have an external $\frac{3}{4}$ inch Acme thread at one end, and a $\frac{1}{2}$ inch internal thread at the tapered end (see Fig. 3-6).

Two she bolts are connected by a $\frac{1}{2}$ inch rod that is threaded to fit into the she bolt. The length of the rod is governed by the thickness of the wall and may be determined by subtracting the required setback on both sides of the form from the wall thickness.

She bolts may be used with walls of any thickness by using the proper inside tie rod. The she bolt and nut washers are reusable. Only the inexpensive inside tie rod remains in the concrete. Other advantages of she bolts are that the ties can be placed after both sides of the form are in place and it can be used for gang forming.

FORMS FOR CONCRETE WALLS 44

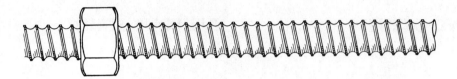

CONTINUOUS COIL THREADED ROD

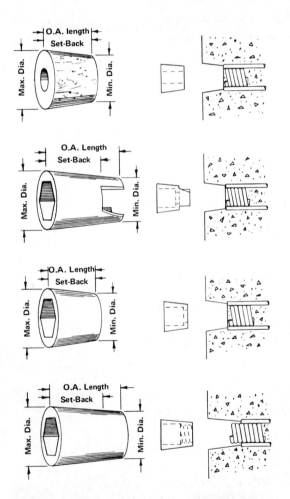

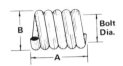

COILS

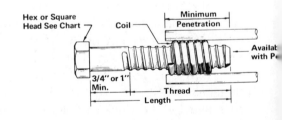

COIL BOLTS

SETBACK CONES

FIG. 3-5. Coil Tie Accessories. Courtesy Dayton Sure-Grip & Shore Co.

FORMWORK MATERIALS 45

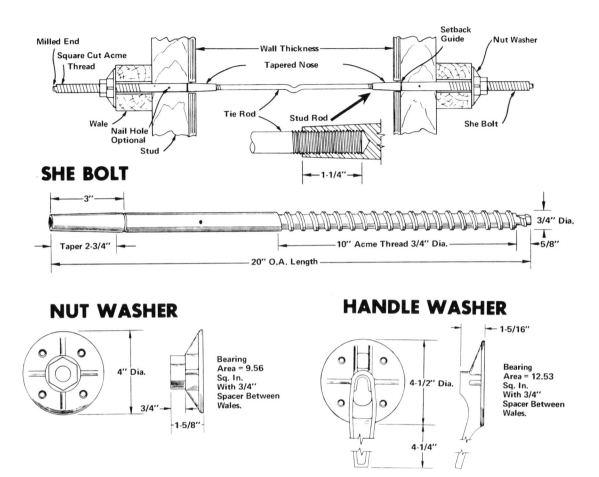

FIG. 3-6. She Bolt Tie. Courtesy Dayton Sure-Grip & Shore Co.

The rod clamp tie (see Fig. 3-7) consists of clamps with set screws, rods of $\frac{1}{4}$ inch, $\frac{3}{8}$ inch, or $\frac{1}{2}$ inch diameter, and a tightening wrench that is placed over the long end of the rod and turned to draw the form together. After the tie is locked with the set screw on the clamp, the wrench is removed. Rod

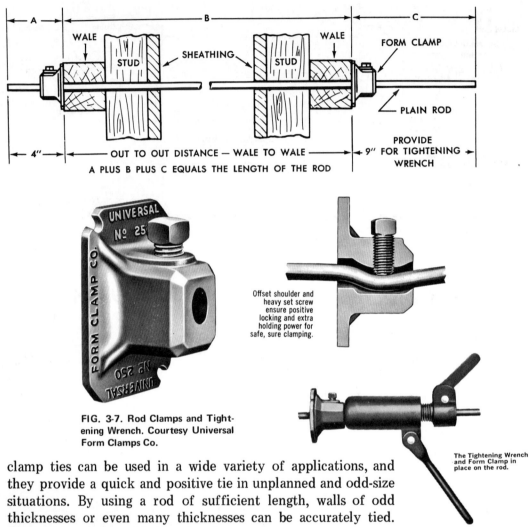

FIG. 3-7. Rod Clamps and Tightening Wrench. Courtesy Universal Form Clamps Co.

clamp ties can be used in a wide variety of applications, and they provide a quick and positive tie in unplanned and odd-size situations. By using a rod of sufficient length, walls of odd thicknesses or even many thicknesses can be accurately tied. Spacers must be provided separately.

Various looped end and flat ties are available for use with the forming systems manufactured by a number of companies. Most of these snap ties serve as both tie and spreader when used with the proper connecting hardware.

PRESSURES ON FORMWORK

Concrete forms must be built strong enough to resist the various forces they are subjected to with an adequate margin of safety. These forces include live loads of men and equipment, dead loads of forms and reinforcing steel, the pressure of fresh concrete, and wind loads. The forces that act on a form depend on where the form is located, and not all forms will be subjected to all the various forces.

Building codes regulate the design and construction of concrete forms and generally accord with the requirements established in ACI347-68, Recommended Practices for Concrete Formwork (see Appendix A). Although the carpenter is not responsible for the design of the form, he should be aware of the forces to which it will be subjected. Quite obviously, men, equipment, wind, and concrete exert pressure on forms, but it is more difficult to realize the effects of lateral concrete pressures on vertical form surfaces.

Most concrete is assumed to weigh 150 pounds per cubic foot. It acts as a liquid and initially exerts a lateral pressure equal to its depth times its density. As it sets, however, it starts to support itself and the lateral pressure is reduced. The rate of set is due, in part, to the temperature, and the lateral pressure on the forms depends on the height of the liquid concrete (rate of pour), the temperature of the concrete, the method of consolidation, and other factors.

For regular concrete with a 4 inch slump, the American Concrete Institute recommends the following formulas for determining lateral pressures on wall forms:

1. ordinary work with internal vibration with the rate of placement 7 feet per hour or less,

$P = 150 + 9000 \frac{R}{T}$ (maximum 2000 lb per sq ft (PSF) or $150h$, whichever is least)

2. ordinary work with internal vibration with the rate of placement over 7 feet per hour,

$P = 150 + \frac{43{,}400}{T} + 2800 \frac{R}{T}$ (maximum 2000 PSF or $150h$, whichever is least)

P = lateral pressure in pounds per square foot

R = rate of pour in feet per hour

T = temperature in degrees Fahrenheit

h = height of fresh concrete above point considered in feet

Table 3-4. Concrete Pressures on Wall Forms*
(in pounds per square foot)

RATE OF POUR (Feet Per Hour)	TEMPERATURE OF CONCRETE (50° F)	(70° F)
1	330	280
2	510	410
3	690	540
4	870	660
5	1050	790
6	1230	920
7	1410	1050
8	1470	1090
9	1520	1130
10	1580	1170
11	1630	1210
12	1690	1250

*Pressure need not exceed $150h$.

The lateral pressures developed in wall forms with concrete at 50° F and 70° F based on the preceeding formulas are tabulated in Table 3-4 but in no case will the actual pressure exceed $150h$.

If pressures on formwork are known, the form sheathing, stud sizes, and tie spacings can be determined. The allowable pressure on plywood sheathing continuous over two or more supports is tabulated in Table 3-5. This table can be used to advantage when designing a form utilizing $\frac{5}{8}$ inch or $\frac{3}{4}$ inch Class 1 Plyform.

Table 3-5. Allowable Pressures (PSF) on Class 1 Plyform
Based on American Plywood Association Data*

SUPPORT SPACING	FACE GRAIN ACROSS SUPPORTS (Plywood Thickness)		FACE GRAIN PARALLEL TO SUPPORTS (Plywood Thickness)	
	5/8"	3/4"	5/8"	3/4"
4"	4070	5010	2250	2780
8"	1370	1740	870	1070
12"	610	770	380	660
16"	290	400	160	330
20"	150	220		170
24"		130		120

*Plywood continuous over two or more supports; deflection limited to $\frac{span}{360}$.

To design a form using plywood and lumber, the following steps would be followed:

1. determine maximum pressure
2. decide on sheathing thickness

3. determine support spacing
4. determine support span (waler spacing) and stud size
5. determine waler span — tie spacing
6. determine tie load and size

EXAMPLE

Design a form 8 feet high to be filled with 70° concrete at a rate of 2 feet per hour.

1. Maximum pressure from Table 3-4 is 410 lb/sq ft. This amount exceeds 150h, which is 300 lb/sq ft. Therefore, 300 lb/sq ft is the design pressure.

2. Sheathing thickness — $\frac{5}{8}''$ and $\frac{3}{4}''$ concrete form grades are most readily available. Assume that $\frac{3}{4}''$ plywood will be used.

3. Determine support spacing. Table 3-5 shows that $\frac{3}{4}''$ plywood supporting 300 lb/sq ft requires studs 16" O.C. NOTE: Stud spacing is generally limited to a multiple of 96" and most often is established at 12", 13.7", 16", 19.2", or 24" O.C.

4. Determine stud span

300 lb/sq ft $\times \frac{16}{12}$ = 400 lb/lin ft.

Table 3-6 shows that 2 by 4 studs have an allowable span of 33".

Table 3-6. Maximum Spans in Inches for Joists or Studs*

EQUIVALENT UNIFORM LOAD (lb/lin ft)	LUMBER GRADE — HEM-FIR NO. 2							
	CONTINUOUS OVER 2 OR 3 SUPPORTS				CONTINUOUS OVER 4 OR MORE SUPPORTS			
	2 × 4	2 × 6	4 × 4	4 × 6	2 × 4	2 × 6	4 × 4	4 × 6
400	33	48	48	73	36	53	56	81
600	27	39	41	59	27	43	45	66
800	22	34	35	51	22	35	39	58
1000	19	29	31	46	19	30	35	51
1200	17	26	29	42	17	27	31	47
1400	15	24	27	39	16	25	27	43
1600	14	23	24	36	15	23	25	39
1800	14	21	22	34	14	22	23	36
2000	13	20	21	33	13	21	21	33

*Adapted from *Plywood For Concrete*, published by the American Plywood Association.

5. Determine waler span. Waler load is equal to concrete pressure times wale spacing in feet.

$300 \times \frac{33}{12} = 825$ lb.

From Table 3-7, double 2 by 4 wales over 4 or more supports could be supported by ties 32" O.C.

Table 3-7. Maximum Spans in Inches for Double Wales*

EQUIVALENT UNIFORM LOAD (lb/lin ft)	LUMBER GRADE — HEM-FIR NO. 2					
	CONTINUOUS OVER 2 OR 3 SUPPORTS			CONTINUOUS OVER 4 OR MORE SUPPORTS		
	2 × 4	2 × 6	2 × 8	2 × 4	2 × 6	2 × 8
800	31	45	60	34	53	69
1000	29	43	56	31	48	63
1200	27	39	51	27	43	57
1400	24	36	47	24	38	51
1600	22	34	44	22	35	46
1800	20	32	42	21	32	43
2000	19	29	39	19	30	40
2200	18	28	37	18	28	38
2400	17	26	35	17	27	36
2600	16	25	33	16	26	34
2800	15	24	32	16	25	33
3000	15	23	31	15	24	31

*Adapted from *Plywood For Concrete*, published by the American Plywood Association.

6. Determine tie load. The tie load is equal to the concrete pressure times the stud span times the waler span in feet.

$300 \times \frac{33}{12} \times \frac{32}{12} = 2200$ lb.

Therefore, 3000 lb. capacity ties could be used.

Complete authoritative information on concrete form design can be found in ACI's *Formwork for Concrete*.

JOB BUILT FORMS

Job built forms are built on the job site from a combination of the various construction materials, such as 2 by 4's, plywood, boards, and hardware. They are not to be confused with patented form panels that are assembled on the job site but manufactured elsewhere.

The typical job built wall form consists of sheathing, studs, plates, wales, ties and spreaders, stakes, and braces (see Fig. 3-8). Some forms have an additional vertical member called a strongback, liner, or aligner.

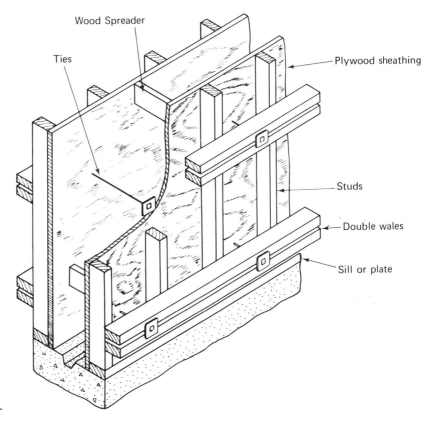

FIG. 3-8. Typical Wall Form.

SHEATHING — Form sheathing is the part of the concrete form that contacts the plastic concrete and gives the concrete its finished shape and surface texture. Many different materials are used individually or in combination for form sheathing, including wood boards (smooth or textured), plywood, hardboard, plastic and fiberglass liners, and steel.

STUDS — Form studs brace the form sheathing. The studs are closely spaced and keep the sheathing aligned and prevent it from bending excessively under the load of wet concrete. The spacing of the studs is governed by the thickness of the sheathing, the temperature and rate of pour of the concrete, and the allowable deflection between studs. Studs are usually made from Douglas Fir, white fir, hemlock, or yellow pine 2 by 4's and 2 by 6's.

PLATES — Plates are not used on all wall forms. In many cases they are used only on one side of the wall form. If used, plates serve to align the bottom of the form wall, and they also provide a means for fastening the form to the footing. Fastening the plate to the footing provides a simple means for maintaining form alignment at the bottom of the form.

WALES — Wales are used to align the wall and to limit the span or unsupported distance between the studs. If wales are more closely spaced, the studs can support a greater load than if they are spaced farther apart. Wales are also called walers or liners.

TIES and SPREADERS — Ties are used to bind the opposite sides of the form together and keep them from moving apart when the form is filled with concrete.

Spreaders keep the form walls properly spaced before the concrete is placed. If twisted wire, tie rods, or other devices are used for ties, spreaders are made from wood 1 by 2, 2 by 2, or other size stock. These spreaders are cut to the exact length required to keep the forms spaced at the finished wall thickness.

Most manufactured ties come equipped with a spreader. This spreader may be a washer held in place by a deformation on the tie, or it may be formed by a stiff tie and compatible connecting hardware.

STRONGBACKS — Strongbacks or liners are used to align and brace the walers. Along with the ties, the strongbacks limit the span of the walers, but their main function is to straighten and align the form vertically. Because they align the form, only straight material should be used for strongbacks.

STAKES and BRACES — Stakes and braces are needed to anchor the form and keep it it proper alignment. Stakes may be made from any material available on the job site, or they may be steel stakes of various lengths to meet job conditions.

Braces are usually made from 2 by 4 or 2 by 6 lumber. They are usually cut to length and nailed to the stakes driven into the ground and to the form. To facilitate aligning the form, a variety of brace clamps with adjustable ends are available (see Fig. 3-9).

Fabricating and Erecting Job Built Forms

A typical wall form similar to that shown in Fig. 3-8 is

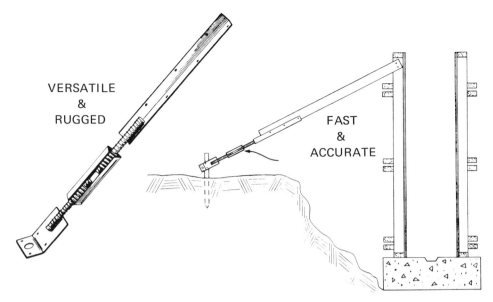

FIG. 3-9. Adjustable Brace Clamps. Courtesy Dayton Sure-Grip & Shore Co.

made up of sheathed stud panels, snap spreader ties, and double 2 by 4 wales with accompanying stakes and braces. The construction of this type of form is discussed in the following paragraphs.

Before any wall forms can be constructed, a thorough study of the plans must be made. After determining the requirements, the form panels of the required size and shape may be built.

These panels will usually consist of 2 by 4 studding and sheathing. The panel size should be standardized whenever possible, and a bench with closely spaced cleats for locating studs should be provided (see Fig. 3-10).

FIG. 3-10. Bench for Building Form Panels.

Studding cut to proper length is placed in the slots on the bench, and the form sheathing is aligned on the studs. The sheathing is nailed to the studs with sufficient nails to secure the assembly. More nails are used if the form will be reused a number of times than if it is to be used only once.

It should be understood that the weight of the concrete will bring the sheathing in contact with the studs, and the nails are needed only to keep the sheathing on the studs before the concrete is placed. However, if the forms are to be reused they will be handled considerably, and the sheathing must be nailed securely to maintain panel size and shape.

While the panels are on the bench, holes should be drilled for the necessary ties, and any hardware such as keyways, shelf inserts, and the like, should also be applied.

It is common practice to start erecting the form at an outside corner. Lines that represent the outside of the wall are established on the concrete footing or floor, and two form panels are fastened together and placed at the corner. These panels are held in place by driving hardened nails through the plate into the concrete. When fastened in place, the face of the form sheathing should be on the established line.

After the panels are fastened in place at the bottom, they are plumbed and braced. Additional form panels for the rest of the outside wall form may now be placed and braces installed as required to maintain form stability.

When a sufficient number of panels have been erected, walers and spreader ties may be installed. This is usually a two-man job, with one workman on the inside of the form passing the tie through holes in the form sheathing. The other man on the outside of the form passes the end of the tie through the waler and installs the wedge tightener or other required fastening hardware.

After an entire wall is erected, it is aligned and permanently braced. A short wall may be aligned by sighting along the top of the form from one end to the other and adjusting the braces as required. On longer walls it is better to sight along a line stretched from one end of the form to the other. For extremely long walls, the transit may be set up and used to align the walls (see Establishing Points on a Line in Chapter 2).

After the outside form is completed, all necessary reinforcing steel is installed. This reinforcing material should not be fastened to the form ties but should be free standing. Reinforcing steel is not installed by carpenters. As a result, the carpenter foreman or job superintendent must be alert to prevent improper tying of the steel.

When a portion of the steel work is complete, erection of the inner wall form may begin. This also is usually a two-man job, with one man adjusting the panel as the other aligns the tie ends with holes in the form panel. As the form panels are placed, the walers are also positioned over the tie rod ends, and the necessary tightening hardware is installed. The form is automatically spaced by the combination tie and spacer, and it is also as straight as the outside wall because it is held parallel to the outside form by the ties.

After completing the inside form panels and installing the required walers, fasteners, and strongbacks, the form is checked for alignment. Additional braces should be installed where the form is out of alignment, and previously installed braces should be adjusted as necessary to align the form completely.

All braces and ties should be checked for security before placing the concrete. They should also be checked periodically while the concrete is being placed to be sure that they are secure and not in danger of breaking or loosening.

Forming Openings in Wall Forms. At various times it is necessary to form openings in a concrete wall for doors, windows, pipes, and other items.

Small openings can be formed by placing fiber, plastic, or metal tubes between the form walls. These tubes are cut to the same thickness as the finished wall and are held in place by wood or plastic rosettes fastened to the form sheathing (see Fig. 3-11).

FIG. 3-11. Tube Inserts. Inserts may be nailed to wall or slab sheathing. Wood or plastic rosettes are also used to fasten inserts.

Slightly larger openings are formed by placing a wood "core" box in the wall. If these openings are less than 8 inches on a side, it is desirable to make the box tapered as shown in Fig. 3-12. The taper makes the box easy to remove after the forms are stripped.

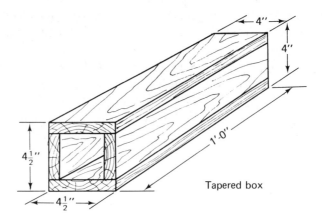

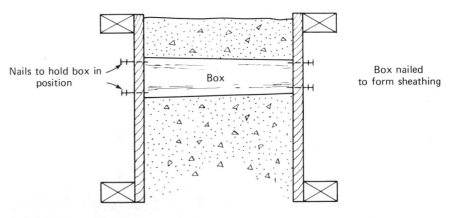

FIG. 3-12. Tapered Core Box.

The location of most openings requiring inserts or core boxes is given on the framing plans. A careful study of the framing plans for the location, size, and number of openings required can result in job economies by avoiding the need to cut holes in the finished concrete.

Window openings can be formed by installing a box of the required size in the form, but in some applications the window frame is placed in the form. This technique yields a completely finished opening when the forms are stripped from the hardened concrete.

A simple window opening box may be made from nominal 2 inch thick lumber. This material is ripped to the exact width

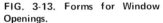

FIG. 3-13. Forms for Window Openings.

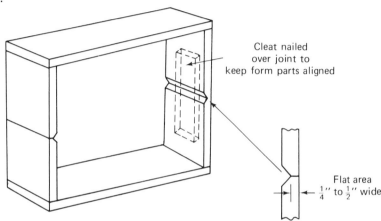

A. Removable window frame

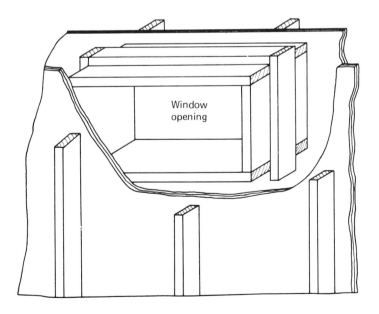

B. Permanently installed window form

required to fill the space between the forms, and in width is equal to the wall thickness. If the wall thickness exceeds the width of the available lumber, two or more strips may be cleated together to make a form of the required width. If the box is to be removed after the concrete has hardened, the oustide dimensions are made to the size of the required rough opening (see Fig. 3-13A).

A construction joint should be made in the sides of the window box (see Fig. 3-13A) so that the box can be removed easily during the stripping operation. By removing the retaining cleats and prying at the joint, the pressure on the side jambs is released and the pieces can be removed. After the side jambs have been removed, the top and bottom jambs are easily released. On forms for small openings that cannot be tapered, a construction joint on all four sides of the form will make it easy to remove the form.

The construction joint can be made by first cutting the pieces to length with square ends. Then the ends can be beveled back so that the completed bevel cut will leave a square cut $\frac{1}{4}$ to $\frac{1}{2}$ inch wide across the width of the piece. This flat area allows the jamb ends to carry some load without crushing, as they would if they were beveled to a sharp edge. The beveled cut provides relief space for swinging the jamb sections out of position.

If the window opening form is left in the concrete wall, it is built so that the inside dimensions are the same as the required opening. The form is anchored to the concrete wall by means of a cleat fastened to the outside of the form (see Fig. 3-13B). This cleat may be made from 1 by 2, 2 by 2, or 2 by 4 stock, depending on the size of the opening. On small openings, a 1 by 2 fastened on two sides of the form may be sufficient, but on large forms in thicker walls a 2 by 4 fastened to all four sides is preferable. Forms may also be anchored to the concrete wall by fastening metal angles to the back of the form. These angles become embedded in the concrete and provide the necessary anchorage.

Window opening forms should always be internally braced with a sufficient number of spreaders (see Fig. 3-14A). These spreaders prevent the weight of the fresh concrete from pushing the form out of shape. The size of the opening and the height of the fresh concrete to be placed around and over the form govern the number of spreaders required.

Spreaders may be made from 2 by 4 or 2 by 6 lumber; but 1 by 4 or 1 by 6 lumber may be used if the spreaders are adequately cross-braced, as shown in Fig. 3-14A. If spreaders are not used, the weight of the fresh concrete will push the form out of alignment, in the manner shown in Fig. 3-14B.

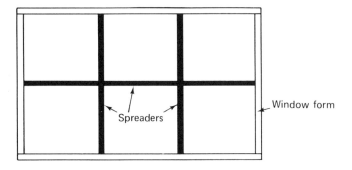

A. Form with spreaders

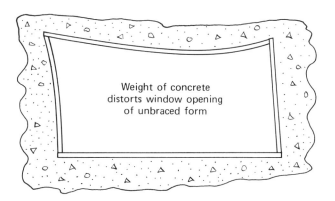

FIG. 3-14. Window Form with Spreaders.

B. Effect of concrete on form without spreaders

The amount of deformation depends on the size of the opening and the amount of concrete placed around the form. Regardless of the size of deformation, the end result is the same; the desired opening is too small, and it must be enlarged by chipping out the hardened concrete after the forms are stripped.

Forms for windows are located in accordance with dimensions on the framing plans. They are fastened to the form sheathing by nailing through the sheathing into the box form with double-headed nails. To prevent the form from moving while the concrete is being placed, it is advisable to nail through the form sheathing into the window form from both the inside and outside wall form.

If window frames are used to form the opening in the wall, the jambs must be the same width as the wall thickness. If the jambs are narrower, they must be built up to wall thickness by adding a filler strip to the frame. This can be done with either wood or steel frames (see Fig. 3-15).

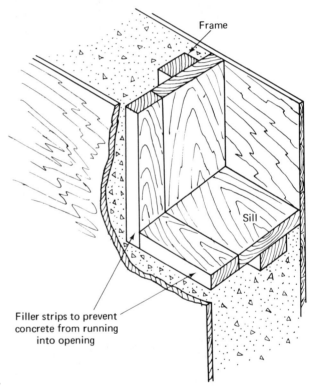

FIG. 3-15. Filler Strip on Frame.

With wood frames, the filler strip is simply nailed to the frame; but if metal frames are used, it is usually necessary to make a composite filler strip, as shown in Fig. 3-16. This type of filler is fastened to the frame with self-tapping screws and is nailed to the form sheathing in the usual manner. The other side of the frame may be fastened to the form with self-tapping screws; but if screw holes in the finished work are unacceptable,

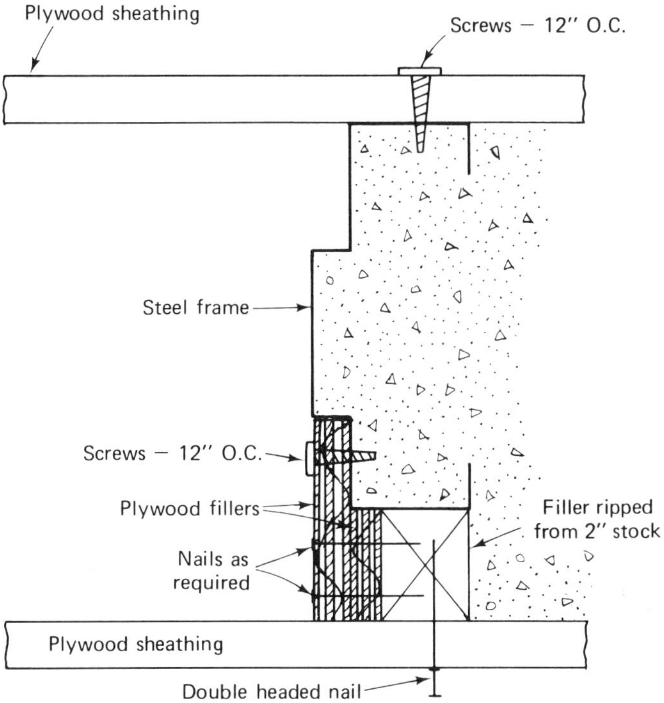

JOB BUILT FORMS 61

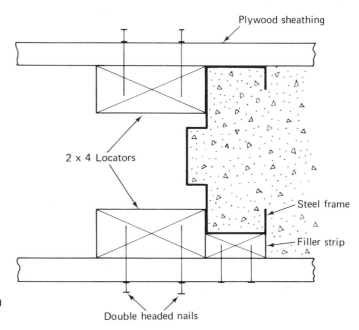

FIG. 3-16. Filler Strip on Steel Frame.

FIG. 3-17. Filler Strip on Steel Frame.

locator blocks may be fastened to the form sheathing (see Fig. 3-17). These blocks are fastened to the sheathing at the inside of the frame and effectively keep the frame in place.

Door openings in concrete walls are formed in a manner similar to that for windows. A door buck (see Fig. 3-18) made from nominal 2 inch thick lumber is fastened to the form sheathing in the proper location. A spreader must be placed at the floor, and at least two additional spreaders should be used between the floor and the header to keep the door buck from becoming deformed under the weight of fresh concrete.

FIG. 3-18. Door Buck.

If the wood door buck is to remain in the opening, a wood strip is attached to the back in the same manner as for window bucks, as shown in Fig. 3-13B. If the buck is to be removed and replaced by a finished lumber jamb, nailing blocks are installed to the back side of the door bucks (see Fig. 3-19). These blocks, because of their triangular shape, are permanently held in the concrete and provide a convenient means for nailing wood door jambs in place.

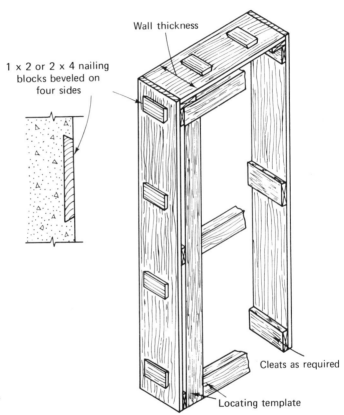

FIG. 3-19. Door Buck with Nailing Blocks.

Metal door bucks are located in the same manner as wood door bucks. The metal spreader at the bottom, which is fabricated as a temporary part of the metal buck, is fastened to the concrete floor with hardened nails or powder-actuated fasteners. After it is fastened to the floor, the buck is plumbed and held in place with locator blocks that are fastened to the form sheathing at the head jamb. Spreaders are installed between the floor and head jamb to keep the frame in alignment.

AllenForm Wall System

A number of manufacturers make ties and connecting hardware for use with lumber and plywood in constructing wall forms. At least one company has developed a system for building wall forms with these materials. The AllenForm system uses 2 by 4 lumber, 4 inch by 8 inch plywood sheets either $\frac{5}{8}$ inch or $\frac{3}{4}$ inch thick, and the AllenForm snap ties and holders (see Fig. 3-20).

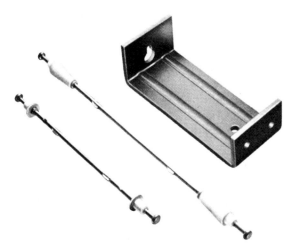

FIG. 3-20. AllenForm Hardware.
Courtesy AllenForm Corp.

The system utilizes panels of one size and requires only one nail in each corner of the plywood panels, four nails per sheet. With this system nailing is kept to a minimum, and when the forms are stripped the materials can be reused on other forms or used for other purposes because nearly all lumber and plywood is used in stock sizes.

The procedure for erecting a form using the AllenForm[1] system is given in the following paragraphs, and the steps are illustrated in Figure 3-21.

Stack 4' by 8' sheets of plywood and square the corners so that several sheets can be drilled at one time. Drill $\frac{1}{2}$" holes on 2' centers, beginning 1' in, and 1' down from any corner (see Step 1). Standard

[1] Procedure courtesy AllenForm Corp.

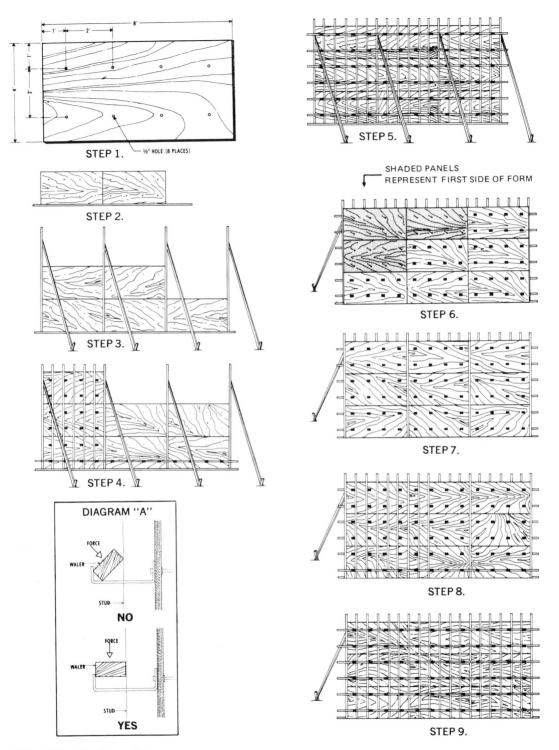

FIG. 3-21. AllenForm System. Courtesy AllenForm Corp.

drilling provides eight holes in each sheet of plywood. Two-foot centers are maintained regardless of how the sheet is placed in the form wall. It is suggested that wherever possible plywood be placed with the 8' side on the horizontal. In this position, with the grain of the face ply running perpendicular to the studs, the plywood offers its greatest resistance to deflection.

FIRST SIDE OF FORM Nail a 2" by 4" plate to the footing, outside and away from the wall line a distance equal to the thickness of the plywood to be used. A plate is required on only one side of the form. Beginning at a corner, set the first sheets of plywood inside the plate. Level the plywood and nail to the plate *at the corners only* (see Step 2).

Position vertical studs lapping the edges of the plywood and nail *at the corners only* so that the vertical joints between the sheets of plywood will be backed up by the studs. Nail plywood only at each corner, *four nails in each sheet* (see Step 3).

From inside the wall insert AllenForm Ply-Ties through the pre-drilled holes and attach AllenForm holders on the outside. Place the bottom waler in position in the bottom row of holders. Lay the waler into the holders in a flat position; do not lay the waler on edge on the bottom of the holder and attempt to twist it into position (see Diagram "A").

Additional vertical studs can now be put into place by sliding them down behind the bottom waler. *STANDARD STUD SPACING $\frac{3}{4}$" Plywood — Studs to be placed on 16" centers — $\frac{5}{8}$" Plywood — Studs to be placed on 12" centers.* No nails are required in these studs; they will be held securely in place by the walers (see Step 4). After all plywood and studs have been positioned, place the remaining walers in position in the AllenForm holders (see Step 5).

SECOND SIDE OF FORM Place pre-drilled plywood over the AllenForm Ply-Ties and secure with AllenForm holders. No studs or walers are required until all plywood has been placed in position (see Step 6).

Position a stud backing up each vertical joint and nail in place at the top and bottom. Follow the same procedure used on the first side of the form in placing the remaining studs and walers in position (see Steps 7, 8, and 9).

MANUFACTURED PANELS

A number of patented manufactured panels are available for building all kinds of wall forms. These panels are made of steel and plywood, all steel, or steel and other materials. Each panel system has its own set of ties, fasteners, connecting hardware, and method of erection.

Gates Forming System

The Gates Cam-Lock forming system utilizes $\frac{3}{4}$ inch plywood in 4 foot by 8 foot or 2 foot by 8 foot sheets, and 2 by 4 or 2 by 6 framing lumber. The plywood sheets are placed vertically in this system with ties set in a 24 inch by 16 inch pattern. No studs are used in the system, but horizontal walers are placed on 16 inch centers and are held against the plywood sheathing by cam-lock brackets. Vertical stiff-backs are placed every 8 feet and held in place with a stiff-back cam.

The Gates forming system uses its own set of ties and connecting hardware. All the hardware, except the tie, is reusable. Additional information on the Gates system can be obtained from either Gates' representatives in major cities or from Gates and Sons directly (see Appendix C).

Some of the extensively used form panels are manufactured by the Symons Manufacturing Co., Universal Form Clamp Co., and Economy Forms Corp. Both Symons and Universal forms utilize a steel frame with plywood sheathing. The Economy form is all steel. Other manufacturers produce forms that are somewhat similar, and they may be used extensively in some localities in place of those mentioned above.

Steel Ply Forms

Symons Steel-Ply Forms (see Fig. 3-22) are made of a heavy steel frame and plastic coated plywood. The standard panels are 24 inches wide and are available in 3, 4, 5, 6, 7, and 8 foot heights. To meet job requirements filler panels 1 inch, $1\frac{1}{2}$ inch, and 2 through 22 inches wide in 2 inch increments are available in the standard panel heights.

The edges of the steel frame come equipped with a dado (groove) that allows the ties to pass through the form. A variety of stock ties are available to meet wall thickness requirements from 4 inches to 48 inches in $\frac{1}{2}$ inch increments, and special length ties are available on request.

Like all forming systems, the Symons system has its own set of connecting hardware, which includes wedge bolts, ties, gang form bolts, and "Z" waler and tie holders (see Fig. 3-23).

FORMS FOR CONCRETE WALLS 68

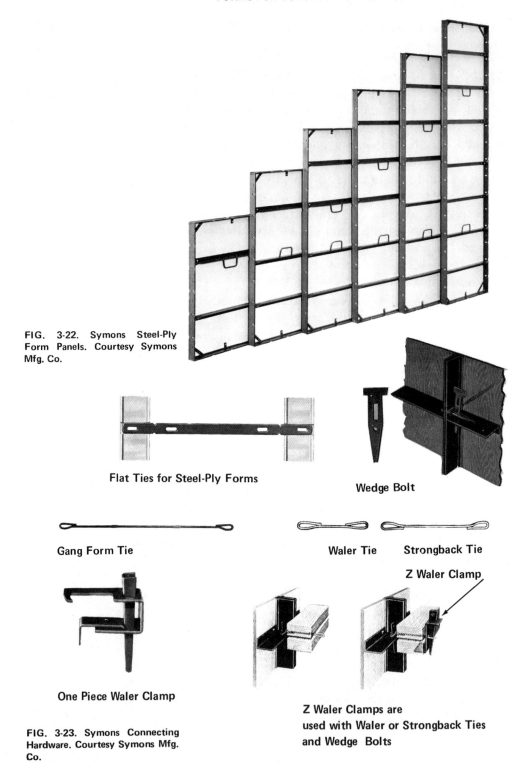

FIG. 3-22. Symons Steel-Ply Form Panels. Courtesy Symons Mfg. Co.

Flat Ties for Steel-Ply Forms

Wedge Bolt

Gang Form Tie

Waler Tie

Strongback Tie

One Piece Waler Clamp

Z Waler Clamp

Z Waler Clamps are used with Waler or Strongback Ties and Wedge Bolts

FIG. 3-23. Symons Connecting Hardware. Courtesy Symons Mfg. Co.

Erecting Steel Ply Forms. For large jobs, the placement of various size form panels is designated on the forming plan. However, on many smaller jobs the forms are laid out by a foreman who has had experience with the sytem. The general procedure followed in building most wall forms with steel ply panels is outlined in the following paragraphs.

The outside of the wall is established and chalk lines are snapped on the footing or floor to represent the outside of the wall. Work usually is begun at an outside corner. The end panels

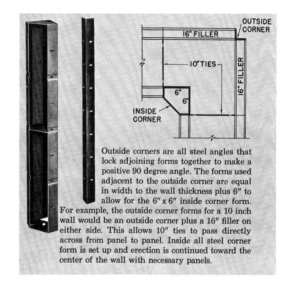

Outside corners are all steel angles that lock adjoining forms together to make a positive 90 degree angle. The forms used adjacent to the outside corner are equal in width to the wall thickness plus 6" to allow for the 6" x 6" inside corner form. For example, the outside corner forms for a 10 inch wall would be an outside corner plus a 16" filler on either side. This allows 10" ties to pass directly across from panel to panel. Inside all steel corner form is set up and erection is continued toward the center of the wall with necessary panels.

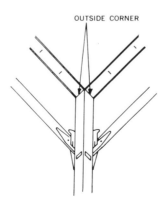

FIG. 3-24. Corner Detail. Wedge bolts face away from corner and are set at the same level on each side of the corner.

meeting at the corner must be 6 inches wider than the wall thickness (see Fig. 3-24A) so that the joints of the inside corner piece will be opposite the joints of the outer end panel. Spreader ties are placed in the joints and locked in place.

When starting the outside corner, wedge bolts are placed through slots in the corner angle adjacent to cross members. The wedge bolt should be pointed away from the outside corner so that the wedge is on the form side of the connection (see Fig. 3-24B). Wedge bolt connections on corners are normally made at the same level as ties in the wall, but they should not be more than 2 feet apart. Wedge bolts in adjacent corner panels should be placed at the same level. Alternating the location of hardware on the outside corners weakens the corner.

After the outside corner is set in place, plumbed, and braced, the second panel is set next to it. Ties are inserted at the dado and the wedge and wedge bolts are installed to hold the tie in place and also to pull the form panels together. Wedges should not be over-tightened, but they should be tight enough to draw the forms together.

The remaining panels are placed along the chalk line and connected in the same manner as the preceding panel. Wedge bolts should be installed from left to right so that they can be driven out with a right-handed hammer blow when stripping the forms (see Fig. 3-25).

When the form wall is completed, the alignment walers and strongbacks, if required, are installed.

In the Symons steel ply system, walers and strongbacks are used for alignment purposes only. The concrete load is carried by the ties and the steel frame of the panel. Therefore, horizontal walers are required on one side of the form only, and are generally placed on the first side of the form to be built. As the walers are placed, the form wall is plumbed, straightened, and braced in position.

In setting up walers, the first step is to thread a waler tie with the panel tie and lock these in position with the wedge bolts. Next, the double 2 by 4 wales are placed over the waler tie, and the "Z" tie holder is slipped over the waler tie and locked in place with the wedge (see Fig. 3-26).

Following the alignment of the first side of the wall form, door bucks, window box-outs, reinforcing steel, and the like, should be placed.

MANUFACTURED PANELS 71

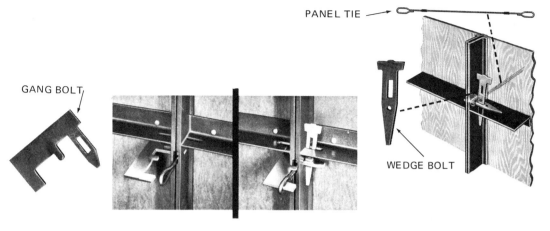

FIG. 3-25. Installing Connecting Hardware.

The inside forms are set in the same manner as the outside forms. Panels opposite each other must be the same width so that the ties line up. As the panels are set, the connecting hardware is installed. Waling or bracing of the inside form generally is not required, but vertical strongbacks may be used on high walls to align the wall vertically.

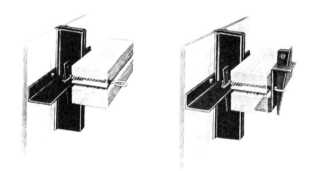

FIG. 3-26. Waler Installation. Courtesy Symons Mfg. Co.

STEP 1—For a waler set-up, the waler tie is threaded with the panel tie and then locked into position by the wedge. Waler ties are reusable.

STEP 2—2x4's are then placed over and under the waler tie. The "Z" tie holder is slipped over waler tie loop against the 2x4's, and the wedge locks the set up.

FORMS FOR CONCRETE WALLS 72

Stripping. In the stripping operation, walers are removed first. Then the wedge bolts and waler ties are removed and dropped into a container so that no parts are lost. Waler ties and wedge bolts are reusable. After the connecting hardware is removed, the panels can be pulled back and removed. Panels should be handled carefully and piled aside for cleaning and reuse (see Cleaning Form Panels, pp. 80-82).

Uni-Form Panels

Uni-Form Panels, manufactured by the Universal Form Clamp Co., are made with a strong steel frame with a $\frac{1}{2}$ inch thick exterior grade plywood for sheathing. Standard panels are 2 feet wide and are made in heights of 1 to 8 feet in 1 foot increments. Closure panels are manufactured in all panel heights in 12 inch and 18 inch widths. Job built fillers of any width can

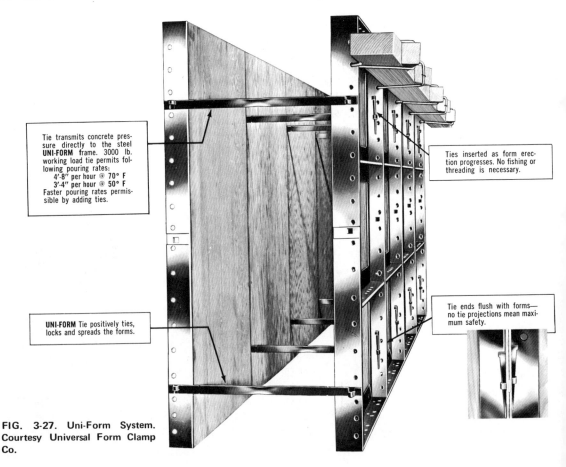

FIG. 3-27. Uni-Form System. Courtesy Universal Form Clamp Co.

MANUFACTURED PANELS 73

be made with $\frac{3}{4}$ inch thick plywood and two Uni-Form angles.

The Uni-Form panel system, like all others, has its own set of connecting hardware. The basic hardware includes ties, tie keys, liner clamps, liner hooks, and panel lock clamps (see Fig. 3-27).

Erecting Uni-Form Panels. Uni-Form panels are erected along snapped chalk lines in the same manner as other panel systems. Starting at an outside corner, form panels are locked together with the outside corner angle and panel lock clamps (see Fig. 3-28A).

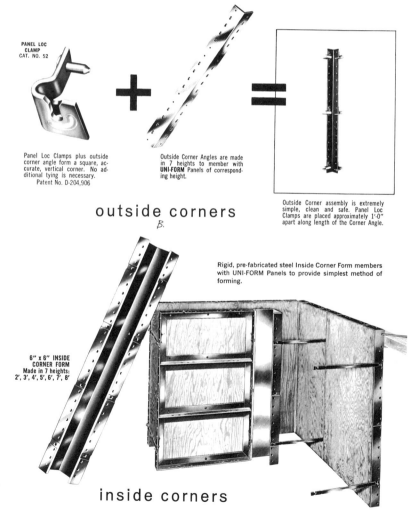

FIG. 3-28. Corner Forming. Courtesy Universal Form Clamp Co.

FORMS FOR CONCRETE WALLS

The end panel must be 6 inches wider than the wall thickness so that the panel joints in the outside form will be opposite panel joints of the inside form when the standard 6 inch steel corner is used (see Fig. 3-28B). By using the all purpose steel filler strip it is possible to build up the steel inside corner to 7, $7\frac{1}{2}$, or 8 inches in width. The width of the outside end panel is always equal to wall thickness plus inside corner size, and the use of the steel fillers makes it possible to form walls of any thickness.

After the corner is set, plumbed, and braced, additional panels are set in place and connected with tie keys. As each panel is put in place, a panel tie is added to the outside end and secured with a tie key. When the next panel is placed, it is locked to the tie and adjoining panel with tie keys (see Fig. 3-29).

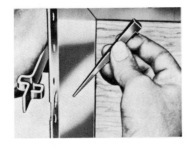 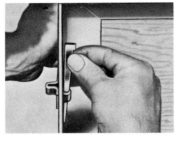 

UNI-FORM Tie Loop is placed in the square tie hole. Tapered Tie Key is set into the Tie loop. This locks the Tie to the Panel.

Bringing the next UNI-FORM Panel into position and inserting a Tie Key in the second loop ties the next form in place. Note that Tie ends are flush with the steel frame.

Simple repetition of these steps on the opposite side closes the wall form.

FIG. 3-29. Assembling Uni-Forms. Courtesy Universal Form Clamp Co.

After all the panels for the outside form are erected, the liners (walers) are installed. Liners may be made from 2 by 4, 4 by 4, 2 by 6, 4 by 6, or 2 by 8 lumber, but the proper size liner clamp must be used.

Liner clamps are fitted over the liners and secured to the holes provided on the frame of the Uni-Form panels. Alignment is accomplished by driving a wood wedge between the liner and liner clamp. A single nail holds the wedges in place. To avoid cutting stock length lumber, the liners may be lapped by using the proper size liner clamp (see Fig. 3-30A).

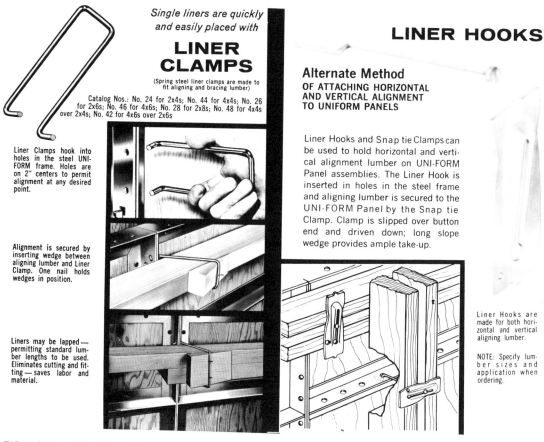

FIG. 3-30. Installing Liners. Courtesy Universal Form Clamp Co.

Liners may be attached with liner hooks and snap tie clamps (see Fig. 3-30B). The liner hooks are inserted in holes of the steel frame, and double liners are placed over the hooks. Standard snap tie clamps are used to tighten the liners against the form panels.

Liners are used for alignment purposes only and are required on only one side of the form. If strongbacks are needed for vertical alignment, they are usually placed on the same side of the form as the liners and fastened with liner hooks and snap tie clamps (see Fig. 3-30B).

After one side of the form is completed, door bucks, window box-outs, reinforcing steel, etc., are placed. Form panels on the second side of the wall are set and held in place by the tie keys. Care should be taken to avoid omitting any tie keys.

Stripping. In the stripping operation, walers and braces are removed first. The tie keys are then removed and placed in a container, such as a gallon pail, for reuse. After the connecting hardware is removed, the panels can be pulled back and removed. Panels should be handled carefully and piled aside for cleaning and reuse (see Cleaning Form Panels, pp. 80-82).

Steel Panel Forms

One of the most commonly used all-steel forms is the EFCO Economy Form manufactured by the Economy Forms Corporation. These forms have a steel frame and steel "sheathing." Panels are manufactured in over 100 standard sizes, and almost any special size can be made on order.

Erecting Steel Forms. After the wall lines have been established, a 2 by 4 base plate should be fastened to the footing with the inside edge of the 2 by 4 along the line. This level base plate makes locating and fastening panels easy, and thereby speeds erection.

The edge of the first regular-sized panel is placed at a distance from the corner equal to the wall thickness plus the corner panel size (see Fig. 3-31) and nailed to the plate. By choosing the right combination of inside corner size and filler panel for the outside form, any wall thickness can be formed.

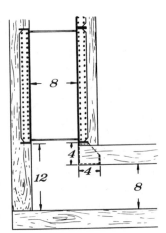

FIG. 3-31. Starting Steel Form Wall. Courtesy Economy Forms Corp.

MANUFACTURED PANELS 77

Additional panels are set alongside the preceding panel and locked in place with the plate clamps. After plate clamping the panel, it should be aligned with the base plate and nailed in position (see Fig. 3-32).

FIG. 3-32. Erection Procedure.
Courtesy Economy Forms Corp.

Erecting first side . . .

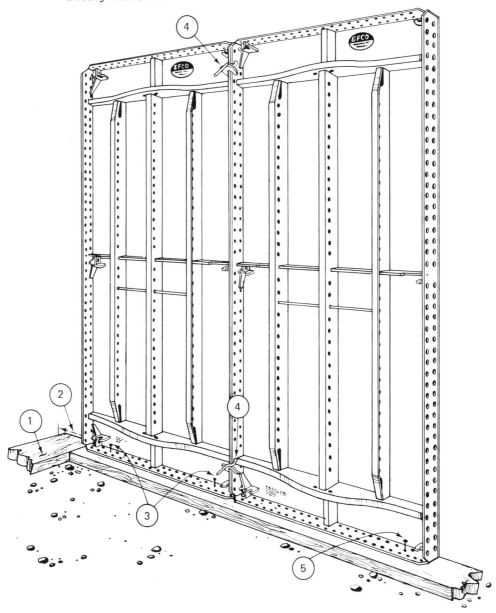

FORMS FOR CONCRETE WALLS 78

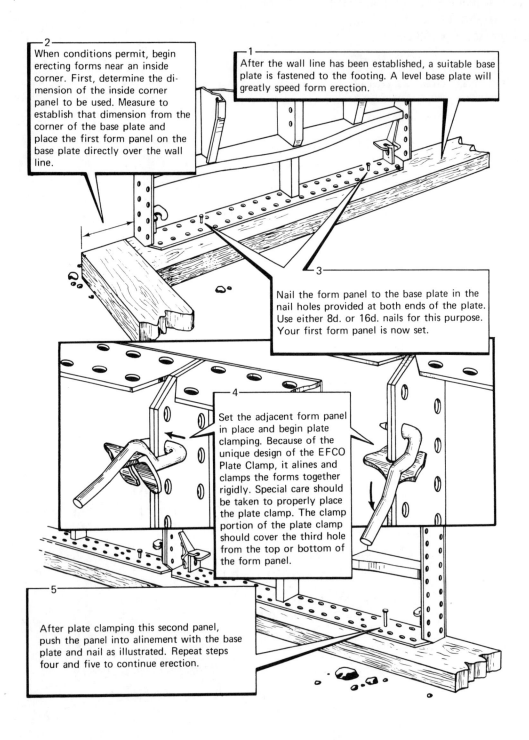

The remaining panels are placed in a similar manner; and after all the panels for one wall are in place, the spreader ties are installed and locked with spreader tie pins (see Fig. 3-33).

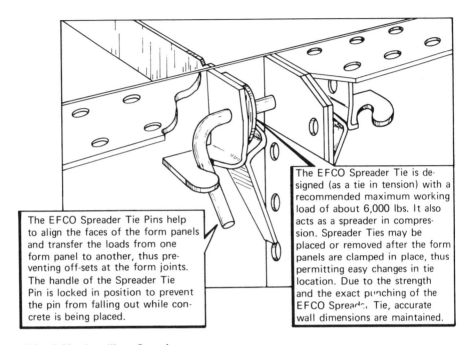

The EFCO Spreader Tie Pins help to align the faces of the form panels and transfer the loads from one form panel to another, thus preventing off-sets at the form joints. The handle of the Spreader Tie Pin is locked in position to prevent the pin from falling out while concrete is being placed.

The EFCO Spreader Tie is designed (as a tie in tension) with a recommended maximum working load of about 6,000 lbs. It also acts as a spreader in compression. Spreader Ties may be placed or removed after the form panels are clamped in place, thus permitting easy changes in tie location. Due to the strength and the exact punching of the EFCO Spreader Tie, accurate wall dimensions are maintained.

FIG. 3-33. Installing Spreader Ties. Courtesy Economy Forms Corp.

The form panel is aligned horizontally and vertically with single 2 by 4, 2 by 6, or 4 by 4 members called aligners. These aligners are held in place with aligner clamps that hook into holes on the panel frame.

Horizontal aligners are usually placed on the outside of the form structure. If the forms are only two panels high, one horizontal aligner is usually sufficient; but if they are three or more panels high, additional rows of aligners are required. The aligners should be lapped at least one form panel to maintain continuous alignment, and one aligner can be lowered at the corners to allow the aligners to overlap and avoid the need for cutting. One aligner clamp must be placed at each vertical form joint in order to align the form (see Fig. 3-34).

Vertical aligners are required if the form set up requires two or more panels vertically. These aligners are usually placed

FORMS FOR CONCRETE WALLS 80

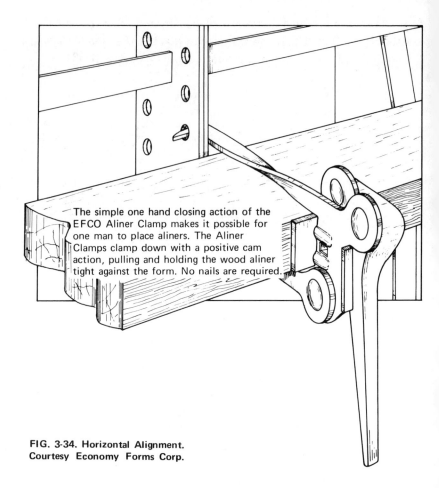

FIG. 3-34. Horizontal Alignment.
Courtesy Economy Forms Corp.

on the inside wall of the form structure at 6 to 8 foot centers. Aligner clamps are required at each joint; and to ensure that the top clamp will be secure, the aligner should be slightly higher than the form (see Fig. 3-35A).

Bracing to keep the forms plumb may be accomplished with adjustable brace clamps, as shown in Fig. 3-35B, or single 2 by 4's may be nailed to stakes and the vertical aligner.

Cleaning Form Panels

After forms have been stripped from the concrete, they should be scraped clean using a putty knife, ice scraper, or other tool. Care should be taken to avoid damaging the form sheathing. The surface of the form may also be cleaned with a wire brush.

FIG. 3-35. Vertical Alignment and Bracing. Courtesy Economy Forms Corp.

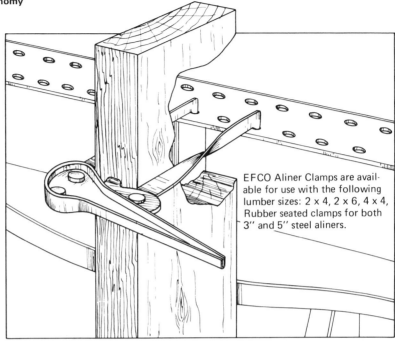

A. Vertical alignment

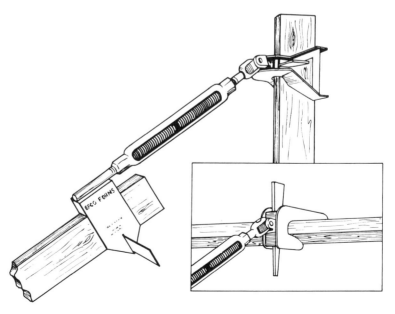

B. Brace adjustment clamp

On large jobs it is usually advisable to use a form cleaning machine, which removes concrete from the side rails and sheathing and applies form oil in one operation. If the forms are to be stored, they should be protected from the weather but not be covered with plastic or other material in a way that prevents air from circulating. Any covering can be used to keep the forms dry; but because form lumber contains moisture, a tight cover may cause the stored forms to become mildewed, and steel forms may rust and collect condensation. Therefore, ventilation around stored forms is as important as protection from the weather.

Gang Forms

A large part of the cost of concrete construction goes for concrete formwork, and any system that can cut down on the cost of forming while still giving satisfactory results is advantageous to the builder and owner.

The steel and plywood panels and the all-steel panels previously discussed can be built into large panels called gangs by using special ties and connecting hardware. These large panels, because of their size and weight, are moved and set with the aid of a crane (see Fig. 3-36).

FIG. 3-36. Gang Form. Courtesy Flour-Utah and Symons Mfg. Co.

One heavy-duty form used for gang forming large flat surfaces is the Symons Superform. The basic Superform unit is a 4 foot by 4 foot frame made of 4 inch chanel side rails and 3 inch chanels for end and intermediate rails (see Fig. 3-37). Filler frames are made in 1 foot, 2 foot, and 3 foot widths with a height of 4 feet. A 4 foot wide by 2 foot high frame is also available.

FIG. 3-37. Superform Frame. Courtesy Symons Mfg. Co.

Two 4 foot by 4 foot frames are bolted to standard 4 foot by 8 foot sheets of plywood, and the 4 foot by 8 foot sections are assembled into gangs by bolting the frames of adjacent panels together (see Fig. 3-38).

FORMS FOR CONCRETE WALLS 84

FIG. 3-38. Assembling Superform. Courtesy Symons Mfg. Co.

Ties are placed 4 feet O.C. at the panel intersections. Therefore, each tie is required to support 16 square feet of form, and ties with a 25,000 pound capacity are used (see Fig. 3-39).

REVIEW QUESTIONS

1. What are the main functions of wall forms?
2. What kinds of lumber may be used for wall forms?
3. What parts of a form may be made from board lumber?

FIG. 3-39. Superform Ties. Courtesy Symons Mfg. Co.

4. How much 1 by 8 T & G boards should be ordered for a wall form 150 feet by 10 feet?
5. What allowance is made for waste when calculating sheathing requirements?
6. List the types of plywood used for form building.
7. List some of the form release agents used on concrete forms.
8. What types of nails are used on concrete formwork?
9. What sizes of wire are most often used for twisted wire ties?
10. What is a snap tie?
11. What is the purpose of wood or plastic cones?
12. When are stainless steel ties preferable?
13. What are coil ties?
14. What are she bolts?
15. What is a rod clamp tie?
16. What forces act against concrete formwork?
17. What does "regular" concrete weigh per cubic foot?
18. Sketch a typical wall form. Label the parts.
19. Outline a method for building form panels.
20. Outline a procedure for erecting wall form panels.
21. How are small openings formed in a concrete wall?
22. How are window and door openings formed in a concrete wall?
23. What precautions should be taken when forming door and window openings?
24. Outline the procedure for building a wall form with AllenForm equipment.
25. Outline a procedure for erecting Symons Steel-Ply Forms.
26. Outline a procedure for erecting Uni-Form panels.
27. Outline a procedure for erecting steel panel forms.
28. How are form panels cleaned?
29. What are gang forms?
30. How are gang forms built and installed?

FORMS FOR CONCRETE COLUMNS
CHAPTER FOUR

Forms for concrete columns may be made from a combination of lumber, plywood, and column hardware, or they may be made from fiber tubes, manufactured form panels, or all-steel column forms.

Because of their comparatively small cross-sectional area, column forms are easily filled with concrete. Fast filling of the form results in a high liquid head of concrete, which in turn places a heavy load on the formwork. The carpenter should realize that this condition exists and be sure to build the forms adequately in accordance with the plans provided.

PRESSURES ON COLUMN FORMS

For regular concrete with a 4 inch slump and internal vibration, the American Concrete Institute recommends the following formula for determining design pressure:

$P = 150 + 9000 \dfrac{R}{T}$ (maximum 3000 lb/sq ft or $150h$, whichever is least)

P = lateral pressure in pounds per square foot

R = rate of pour in feet per hour

T = temperature of concrete in degrees Fahrenheit

h = height of fresh concrete above point considered in feet

Lateral concrete pressures on column forms based on the foregoing formula are tabulated in Table 4-1.

Table 4-1. Concrete Pressures on Column Forms*
            (in pounds per sq ft)

RATE OF POUR (Feet Per Hour)	TEMPERATURE OF CONCRETE 50° F	TEMPERATURE OF CONCRETE 70° F
1	330	280
2	510	410
3	690	540
4	870	660
5	1050	790
6	1230	920
7	1410	1050
8	1590	1180
9	1770	1310
10	1950	1440
11	2130	1560
12	2310	1690

*Maximum pressure is 3000 lbs/sq ft or 150h whichever is least.

Whenever possible, columns are formed and the concrete is placed before work on the floor slab commences. This procedure utilizes the stiffness of the hardened columns to help stabilize the floor slab formwork.

WOOD COLUMN FORMS

One of the oldest methods of building column forms is with boards running vertically and cleated together to make a panel (see Fig. 4-1). Panels are made up in such a manner so that the end panels overlap the side panels. As illustrated in Fig. 4-1, the end panels are made $1\frac{1}{2}$ inch wider than column size to allow for sheathing thickness on the side panels, and the cleat ends are flush with the panel edge.

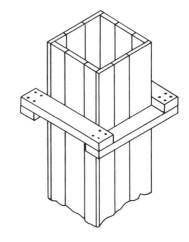

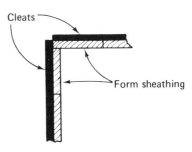

FIG. 4-1. Column Formed with Boards.

Corner detail

The width of the side panels is the same as the column width, but the cleats are allowed to project $\frac{3}{4}$ inch on each side so that they overlap with the end forms when the panels are assembled. Cleats on side and end panels should be set at the same height so that the placing of column clamps will not be hampered. The entire column form assembly is held together with some type of yoke or column clamp. One of the most commonly used column clamps is illustrated in Fig. 4-5.

Plywood and Lumber Column Forms

Most job built column forms are made with plywood sheathing, either $\frac{5}{8}$ inch or $\frac{3}{4}$ inch thick with vertical 2 by 4 stiffeners placed at the edges and at intermediate locations as required by the form plans (see Fig. 4-2).

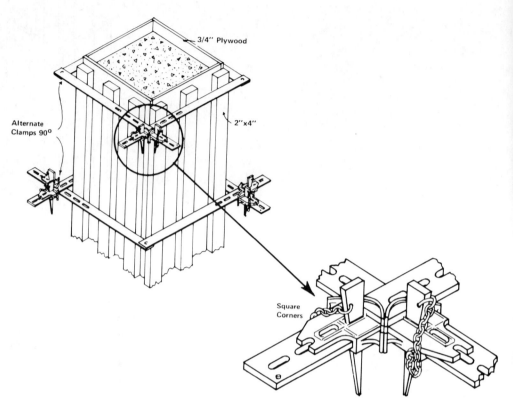

FIG. 4-2. Plywood Column Form with 2 by 4 Stiffeners. Courtesy Dayton Sure-Grip and Shore Co.

Rectangular forms are built with two panels cut to the exact width of the column. The 2 by 4 stiffeners are placed so that they project approximately $1\frac{1}{2}$ inch over the edge of the sheathing. Sheathing for the other two sides is cut $1\frac{1}{2}$ inch wider than the column size to allow for the overlap of sheathing thickness, and the 2 by 4 stiffeners are held flush with the edge (see Fig. 4-3).

When the forms are erected, they are usually first tacked together with double-headed nails driven through the overlapping 2 by 4 stiffeners. Column clamps are placed over the assembly to hold it in alignment (see Fig. 4-5) and to brace it against the concrete load.

WOOD COLUMN FORMS 91

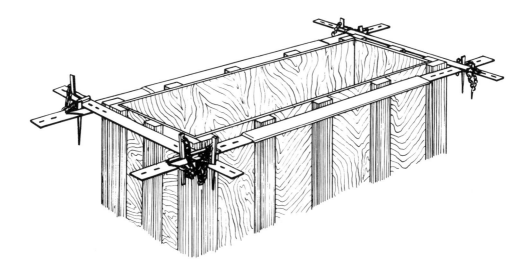

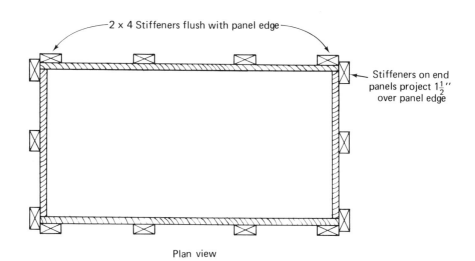

Plan view

FIG. 4-3. Rectangular Column Form.

Forms for square columns are made with all four sides overlapping and of the same width (see Fig. 4-4). When using $\frac{3}{4}$ inch plywood, the stiffeners project $\frac{3}{4}$ inch over the edge of the sheathing, as in Fig. 4-4; but if $\frac{5}{8}$ inch plywood sheathing is used, the edge projection is only $\frac{5}{8}$ inch. When erected, the

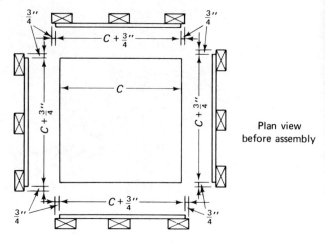

C = Column dimension

Note: When $\frac{5}{8}''$ plywood is used all $\frac{3}{4}''$ dimensions are changed to $\frac{5}{8}''$.

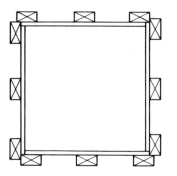

FIG. 4-4. Square Column Form.

forms are usually held together with adjustable column clamps (see Fig. 4-5).

Form Installation. Column forms must be accurately located. Many times the column location will be given by the intersection of centerlines. These lines are impossible to see after the form is placed, and they are difficult to see after the reinforcing steel is in place. Therefore, a template made to the outside dimensions of the form is located and secured to the floor or footing (see Fig. 4-6). The template is made from 2 by 4 material that is overlapped and nailed together with double-headed nails.

The template is located in accordance with the established centerlines and fastened with hardened nails. Before the form

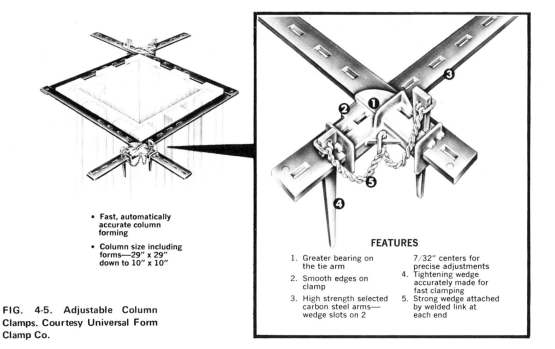

- Fast, automatically accurate column forming
- Column size including forms—29" x 29" down to 10" x 10"

FEATURES

1. Greater bearing on the tie arm
2. Smooth edges on clamp
3. High strength selected carbon steel arms—wedge slots on 7/32" centers for precise adjustments
4. Tightening wedge accurately made for fast clamping
5. Strong wedge attached by welded link at each end

FIG. 4-5. Adjustable Column Clamps. Courtesy Universal Form Clamp Co.

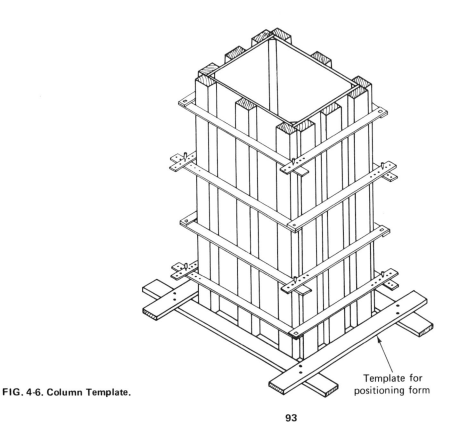

Template for positioning form

FIG. 4-6. Column Template.

sections are placed, they should be thoroughly oiled to prevent or lessen the tendency of the form to stick to the concrete and thereby to simplify stripping. The sections of the column form are placed inside the template and temporarily fastened with double-headed nails.

Column clamps are applied in accordance with the spacing schedule supplied with the plans. The maximum pressure will be at the bottom of the form. Therefore, form clamps should be more closely spaced near the bottom than at the center or near the top. Table 4-2 gives column clamp spacing for typical columns.

Table 4.2. Recommended Column Clamp Spacing

CLAMP SIZE	36"			Distance from TOP
SPAN	14"/20"	25"	29"	
				2"
	24	30	30	1' – 2'
	24	24	22	3' – 4'
	24	18	18	5' – 6'
	20	16	18	7'
		12	12	8'
	20	12	12	9'
	6	6	6	10'

SPACING OF COLUMN CLAMPS IN INCHES

Based on: 1. Ten foot rate of pour; 2. Pure liquid head of 150 lbs. per square foot of accumulated depth; 3. Deflection of: length of the stand : 270.
Maximum span governs spacing if columns are rectangular.

The column form must be plumbed and braced in two directions. Plumbing can be accomplished by hanging a plumb bob from the top of the form and adjusting the braces as required. The plumb bob must be shielded from the wind so that it will not be blown from a true vertical position.

To avoid problems caused by wind, many contractors prefer to plumb the column forms with the aid of a transit (see Chapter 2). On small jobs it may be permissible to plumb the forms with a level and a straight edge (see Fig. 4-7).

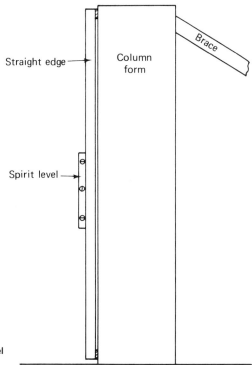

FIG. 4-7. Plumbing with a Level and Straight Edge.

Bracing is usually accomplished by nailing 2 by 4's near the top of the form and to a stake driven into the ground. These braces should make an angle of approximately 45° with the horizontal and should be placed on two adjacent sides. To aid in adjusting the braces, some contractors prefer to use the adjustable brace clamp produced by various manufacturers (see Fig. 4-8).

If there are a number of columns in a row, the end columns are braced in two directions with diagonal braces, and the central columns are held with horizontal spacers placed near the top to hold them in one direction. Diagonal braces are needed only in the other direction.

FORMS FOR CONCRETE COLUMNS 96

FIG. 4-8. Adjustable Braces.

Simple Column Forms

Forms for small columns may be made with plywood sheathing and 2 by 4 cleats (see Fig. 4-9). This type of form is used where pressures are not great and if the work must be done quickly.

With this type of form sheathing, two sides are cut to the exact width of the finished column. The other two sides are made 9 inches wider to accommodate the 2 by 4 cleats at each

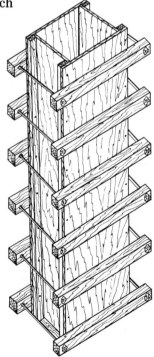

FIG. 4-9. Simple Column Form.
Courtesy Symons Mfg. Co.

edge and to allow $\frac{3}{4}$ inch for the sheathing thickness on each side.

When the form is put together, the plain sheathing panels act as spreaders for the side panels, and the 2 by 4 cleats act as supports for these panels. The entire assembly is held together with adjustable steel column clamps.

Column forms may be secured with steel banding in place of column clamps, and in some cases they are held together with wood and steel yokes (see Fig. 4-10).

FIG. 4-10. Alternate Methods of Clamping Columns.

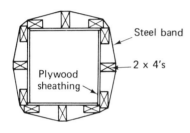

A. Clamping with steel band

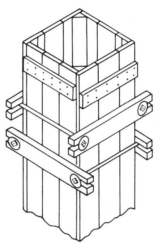

B. Clamping with wood and steel yokes

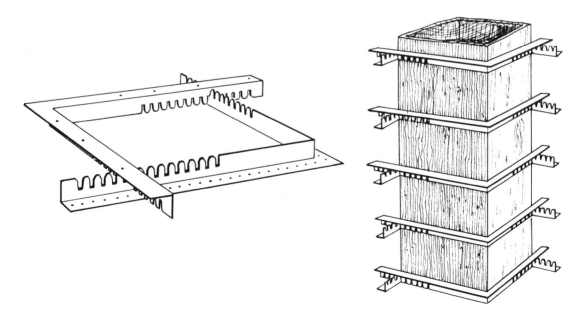

C. Econ-o-clamps — Courtesy Sonoco Products Co.

FIBER FORMS

Fiber forms are used for forming round and obround columns. They are available in standard lengths to 18 feet and may be ordered in longer lengths. The inside diameter sizes range from 6 inches to 48 inches (see Table 4-3).

Table 4.3. Fiber Form Data. Courtesy Sunoco Products Co.

SIZES AND WEIGHTS

Regular (Standard Length—18')							
Inside Diameter	Wt. Per Ft.	Inside Diameter	Wt. Per Ft.	Inside Diameter	Wt. Per Ft.	Inside Diameter	Wt. Per Ft.
6"	1.2 lbs.	18"	5.2 lbs.	28"	10.0 lbs.	38"	16.2 lbs.
8"	1.6 lbs.	20"	7.2 lbs.	30"	11.3 lbs.	40"	17.0 lbs.
10"	2.2 lbs.	22"	7.9 lbs.	32"	12.1 lbs.	42"	17.9 lbs.
12"	2.6 lbs.	24"	8.6 lbs.	34"	12.8 lbs.	44"	19.7 lbs.
14"	3.4 lbs.	26"	9.3 lbs.	36"	13.9 lbs.	46"	20.6 lbs.
16"	4.6 lbs.					48"	22.6 lbs.

Fiber forms are positioned in a template in the same manner as job built forms (see Fig. 4-11). A 2 by 4 bracing frame or collar is placed around the form near the top to aid in attaching the braces used to keep the form plumb.

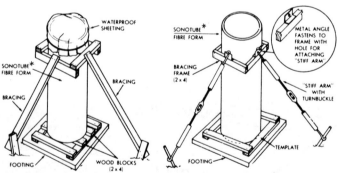

FIG. 4-11. Locating and Bracing Fiber Forms. Courtesy Sonoco Products Co.

If there is a delay in placing concrete after the fiber forms have been placed in position, the form should be protected against rain and snow by covering the top with a waterproof sheathing and by keeping snow and rain from accumulating around the base.

Fiber forms should be stripped as soon as possible after the concrete has set, and in most cases the easiest and fastest stripping can be done if the form is stripped within ten days of pouring.

The usual procedure for stripping fiber forms is to first set a portable saw to a depth equal to the thickness of the form and make two vertical cuts for the entire length of the form. The two halves may then be removed from the concrete. If the concrete requires additional curing, the form halves may be replaced on the column and wired in place. In some cases the form is replaced after stripping to protect the column during construction.

Fiber forms can be cut on the job with ordinary wood-cutting tools. Cutouts may be made for outlet boxes, beam pockets, and other requirements in the same manner as for wood forms.

Fiber tubes may be used for form pilasters by ripping the tube in half lengthwise. The tube is attached to the wall form sheathing and held in place by walers and blocking (see Fig. 4-12).

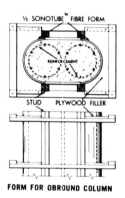

FORM FOR OBROUND COLUMN

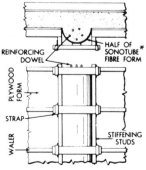

FIG. 4-12. Forming Details.
Courtesy Sonoco Products Co.

FORM FOR PILASTER (column integral with wall)

Obround columns may be formed by cutting a tube in half lengthwise and placing plywood sheathing between the two halves, as shown in Fig. 4-12. The edges of the tube halves and plywood sheathing are supported by 2 by 4 studding to keep the form straight. The round ends of the form are supported by a plywood filler cut to fit the outside of the form. The entire structure may be held together either with wood yokes made on the job or by adjustable column clamps placed over the end fillers and studs.

Details of other commonly encountered forming situations are illustrated in Fig. 4-13.

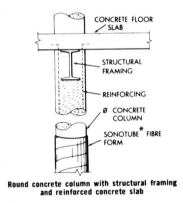

Round concrete column with structural framing and reinforced concrete slab

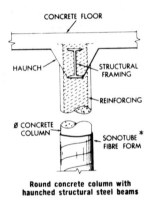

Round concrete column with haunched structural steel beams

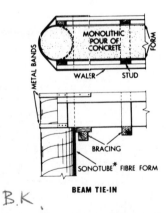

BEAM TIE-IN

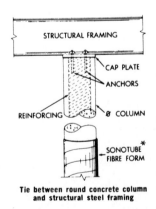

Tie between round concrete column and structural steel framing

FIG. 4-13. Forming Details.
Courtesy Sonoco Products Co.

PANEL SYSTEMS FOR COLUMNS

Plywood panels with steel frames of the type used in wall forms can be fabricated into column forms. The panel width used for the form is the same as the width of the finished column. The panels are joined at the outside corners with corner angles and fastening hardware. Symons Steel-Ply panels are fastened together with outside corner angles and wedge bolts (see Fig. 4-14). The wedge bolts are placed on 12 inch centers at each horizontal stiffener. The only exception is at the joint between the corner angles, where the wedge bolts are placed 6 inches from the end of the angle, and no wedge bolts are placed at the horizontal stiffener in line with the splice between the angles.

FIG. 4-14. Steel Ply Column Form. Courtesy Power Construction Co. and Symons Mfg. Co.

Panels may be stacked two or more high to attain column height. The horizontal joint between panels is locked in alignment with wedge bolts. Joints of corner angles and steel-ply panels are staggered to strengthen the column form and to help maintain panel alignment.

A neoprene rubber chamfer strip is available for use with steel-ply forms. It is made to clamp in place between the corner angle and steel panel frame, yields a smooth uniform chamfer (beveled corner), and can be reused indefinitely.

Uni-Form panels are connected with outside corner angles and panel lock clamps (see Fig. 3-28) and are built into column forms in much the same manner as steel ply forms. Both systems may be gang formed.

ALL STEEL COLUMN FORMS

Round steel column forms are manufactured in diameters from 12 inches to 96 inches and in lengths of 1 foot to 12 feet. The smaller diameters are supplied in 180° sections, and the larger diameters are furnished in 90° sections to facilitate handling.

Round steel column forms may be used with flat steel panels to form obround columns of any size (see Fig. 4-15).

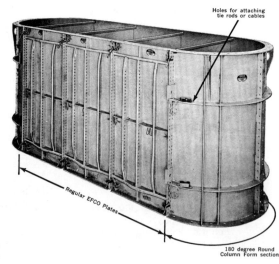

FIG. 4-15. Obround Steel Form.
Courtesy Economy Forms Corp.

These forms are fastened together with the standard connecting hardware that is used on square or rectangular column form panels manufactured by the same company. Therefore, it is advantageous for a contractor to use forms made by one company on a given job rather than to use several different brands.

Columns One Panel Wide...

When the width of the column can be formed with one EFCO Form Panel, internal spreader ties are not required. External ties are required at about 2' centers. The external ties are fastened to the outside tie angle corner with EFCO Plate Clamps. This set-up is useable on columns up to 36" square and has a maximum pouring pressure of 1200 pounds per square foot. Note that the horizontal joints of the EFCO Form Panels and the outside tie angle corners are staggered. Plate clamps should be located as illustrated.

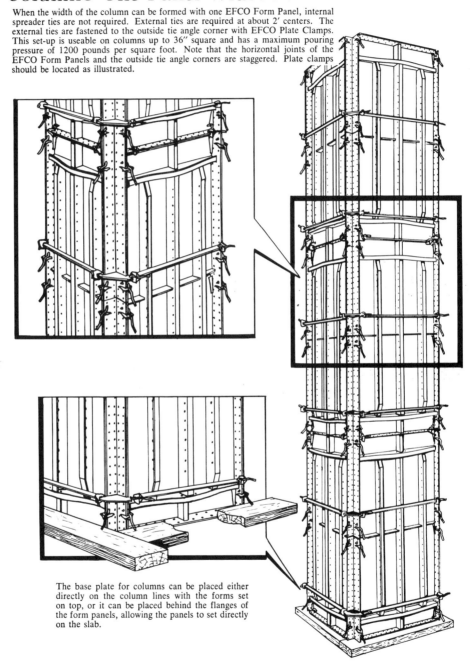

The base plate for columns can be placed either directly on the column lines with the forms set on top, or it can be placed behind the flanges of the form panels, allowing the panels to set directly on the slab.

**FIG. 4-16. Steel Column Form.
Courtesy Economy Forms Corp.**

Bracing...

When a series of columns are formed along the same line, a horizontal aliner can be used to simplify the alining and bracing as illustrated.

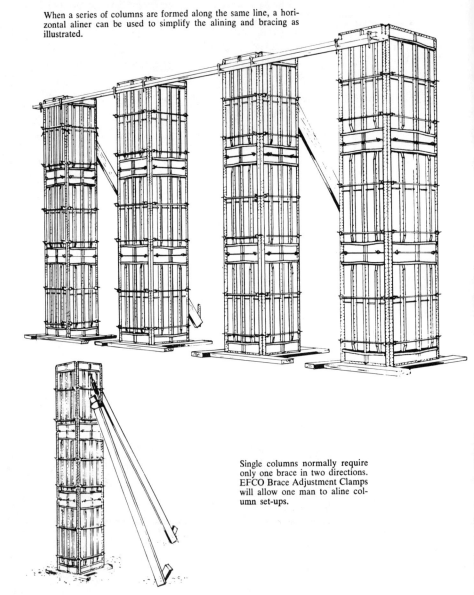

Single columns normally require only one brace in two directions. EFCO Brace Adjustment Clamps will allow one man to aline column set-ups.

FIG. 4-17. Bracing. Courtesy Economy Forms Corp.

When using all-steel panels to form square or rectangular columns, it is best to use panels the same width as the column dimension. This allows the form to be built without internal ties and avoids the need for horizontal aligners. Standard hardware for steel column forms include tie angle corners, ties, and plate clamps.

Tie angle corners are fastened to the form panels with plate clamps. The angle corners are available in 24 inch and 48 inch lengths, and the joints between the form panels and angles should be staggered (see Fig. 4-16). Plate clamps should be placed at each end of the angles and approximately 24 inches apart along the entire column length. The form may be set on top of a 2 by 4 base, or it may be set directly on the floor by making the template large enough to fit the outside of the form (see Fig. 4-16).

All column forms must be braced to maintain alignment. Guy wires or lumber braces can be used for bracing. If guy wires are used, they must run in all four directions because they operate only under tension. Lumber braces work under tension and compression and are needed in only two directions to maintain alignment.

If a number of columns are set in a row, they can be held in alignment with horizontal spreaders and a minimum of diagonal braces; but individual columns must have two braces 90° apart to maintain alignment (see Fig. 4-17).

REVIEW QUESTIONS

1. What materials may be used to build forms for concrete columns?
2. Why are column forms often subjected to high lateral pressures?
3. Why is it preferable to have concrete columns placed before starting work on the concrete floor slabs?
4. How are boards in column forms held together?
5. What thickness of plywood is most often used for column forms?

6. Make a plan view sketch showing the make up of a square column form made from plywood and 2 by 4's.
7. How are templates used with column forms?
8. Why are column clamps spaced more closely near the bottom of the form?
9. How may column forms be plumbed?
10. Why must column forms be braced in at least two directions?
11. How are fiber forms secured in place?
12. What precautions should be taken when using fiber forms?
13. How are fiber forms stripped from the finished concrete?
14. What recommendations should be followed when using patented form panels for column forms?

FORMS FOR CONCRETE BEAMS AND GIRDERS
CHAPTER FIVE

The concrete forms required for beams and girders are nearly identical in construction and contain the same basic parts. Therefore, the following discussion on beam forms applies as well to building forms for girders.

Beam forms are supported at the proper elevation by some type of shoring. The bottom of the form is logically called a beam bottom, the sides are called beam sides, and the remaining parts are given such names as kicker, ledger, waler, and tie (see Fig. 5-1).

SHORING BEAM FORMS

Shoring for beam forms may be single post, double post, or tubular steel. Single post shores may be job built or they may be purchased from a manufacturer of forming equipment. Double post shores are usually job built, and tubular steel shores are available from a number of manufacturers on either a rental or purchase basis.

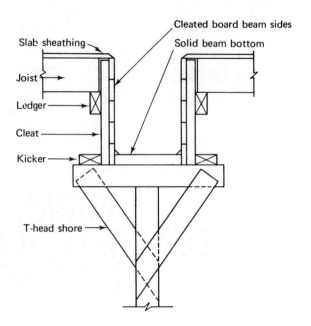

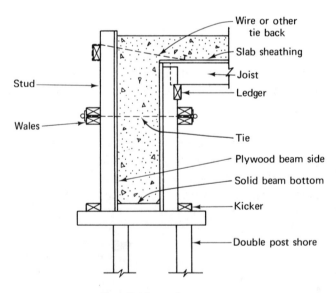

FIG. 5-1. Typical Beam Forms.

Job built single post shores are usually built from 4 by 4's with a T-head or an L-head (see Fig. 5-2). They may be fitted with hardware for making adjustments in length, or they may be cut to within 1 inch of the required length, and the final

108

SHORING BEAM FORMS 109

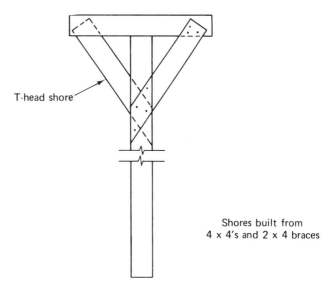

T-head shore

Shores built from
4 x 4's and 2 x 4 braces

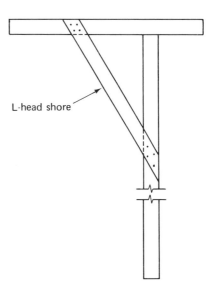

L-head shore

FIG. 5-2. Job Built Shores.

adjustment may be made with pairs of wedges placed under each shore.

Manufactured single post shores have some type of jack or jack screw built into them for adjusting to final length (see

FORMS FOR CONCRETE BEAMS AND GIRDERS 110

Fig. 5-3A). Tubular shoring is manufactured in various sizes and types. Frames are connected in pairs with diagonal braces to form a tower (see Fig. 5-3B). Towers may be stacked by using the proper connecting hardware. Base plates, adjusting screws, and other hardware are available for tubular steel shoring to provide a means for anchoring and adjusting the shores and formwork.

The shoring supports not only the beam form but also workmen, equipment, reinforcing steel, and concrete. Therefore, it must be built sturdily and in accordance with the shoring layout. The shoring layout should be made in accordance with good practice and the safety recommendations of the Scaffolding and Shoring Institute (see Appendix B).

FIG. 5-3A. Single Post Shore. Courtesy Symons Mfg. Co.

SHORING BEAM FORMS 111

5' x 5'

5' x 30"

ADJUSTABLE SCREW JACKS

Adjustable Screw Jacks can be used two ways: for leveling scaffolding on uneven surfaces, and for shoring when used at the bottom or top of the scaffold.

BASE PLATES

U-HEAD

U-Head or Shorehead is used in conjunction with screw jacks for shoring.

Base plates can be used with or without adjusting screw jacks.

FIG. 5-3B. Tubular Frames and Accessories. Courtesy Symons Mfg. Co.

Mudsills

Mudsills of the size specified by the shoring plan should be placed on soil that has been leveled and compacted as required. Sills should be made from sound and rigid material that will be capable of carrying the imposed loads.

Precautions should be taken to avoid placing sills on frozen ground or in areas where rain water runoff can reduce the soil

FORMS FOR CONCRETE BEAMS AND GIRDERS

Shore
Wedges for adjusting length of shore
2 x 10 Sill

Typical mudsill
good soil bearing

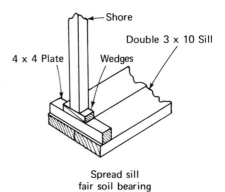

Spread sill
fair soil bearing

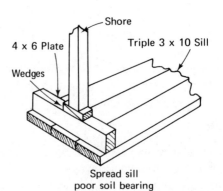

Spread sill
poor soil bearing

Note: Similar mud sills are used under tubular frames and single post steel shores

FIG. 5-4. Mudsills.

load-carrying capacity. If two or more planks are required to make a mudsill of sufficient width, a 4 by 4, 4 by 6, or other short timber should be placed across the mudsill to distribute the shoring load (see Fig. 5-4).

Shore Spacing and Erection

The location of the shores should be marked off on the mudsill. This is generally done by starting at a column and marking the specified center-to-center distance directly on the mudsill. The shores can then be placed at the marks in preparation for erection.

In addition to marking shore location on the mudsill, it may be advantageous to mark a 2 by 4 identical to the mudsill. The 2 by 4 can then be used as a temporary spacer or remain in place at the upper end of the shore.

The individual shores should be adjusted or cut to the approximate length required before erecting. As the end shore in each beam span is erected, it should be adjusted to the established grade and braced vertically. The lower end should be nailed to the mudsill to keep the shore from moving, and the 2 by 4 stay can be fastened in place at the upper end of the shore to provide a means for spacing and vertically aligning intermediate shores.

Intermediate shores can be adjusted to proper length after they are erected by bringing them up to a line stretched across the top of the end shore heads. If camber is required in the beam form, it can be attained by checking the grade of the beam bottom with a transit. The individual shores are then adjusted as required to meet planned specifications.

Bracing

All shoring should be adequately braced. Bracing is required in the longitudinal, transverse, and diagonal directions. Adequate bracing takes time and material, but inadequate bracing is wasteful and dangerous because it leads to form breakdown.

Beam Bottoms

Beam Bottoms are made the width of the finished beam. They may be made up from solid lumber either 1 or 2 inches thick, but most beam forms are made with plywood sheathing supported by 2 by 4 stiffeners that run the full length of the beam (see Fig. 5-5).

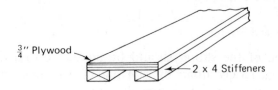

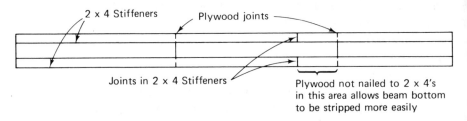

View of underside of beam bottom

FIG. 5-5. Beam Bottoms.

The beam bottom should be fabricated in a way that will facilitate stripping from the hardened concrete. This can be accomplished by making the beam bottom about 2 inches shorter than the distance between the columns and putting a filler or wrecking strip at one end of the beam bottom. If the filler strip is damaged while the form is being removed, it can be easily and inexpensively replaced.

Another way of easing form removal is to make the beam bottom in two sections with a joint 8 feet to 12 feet from one end. The beam bottom is cut to length about $\frac{1}{8}$ inch shorter than the distance between the columns. When the beam bottom is placed, care must be taken to support the joint between the form sections. This is accomplished best by placing the joint directly over a shore head.

The beam bottom is aligned along a string stretched from column to column, and it may be held in place by tacking it to the shoring.

Beam Sides

Beam sides may be made from board sheathing and 2 by 4 studding, or they may be made up with plywood sheathing and 2 by 4 studding. In some cases patented form panels may be used for beam sides.

If beam sides are made on the job, the overall height of the panel is established by determining the distance from the top of the shore head to the bottom of the floor slab sheathing (see Fig. 5-6). Studs for the beam sides are cut approximately $\frac{1}{8}$ inch shorter than the finished height of the beam side so that small irregularities in length will not interfere with the slab sheathing.

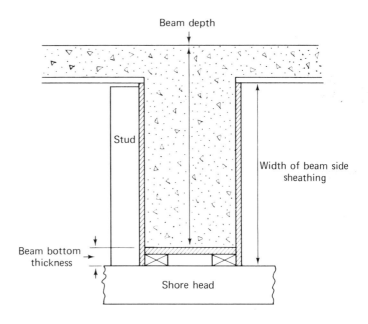

FIG. 5-6. Beam Sides.

Width of beam side = beam depth <u>plus</u> beam bottom thickness <u>minus</u> slab sheathing thickness

Plywood sheathing is ripped to the width required for the beam side and fastened to the studs with enough nails to maintain panel integrity — that is, size and shape. In nailing the sheathing to the studs, care should be taken that no stud ends project over the top edge of the sheathing. Any projection beyond the edge of the beam side sheathing will interfere with the placement of the slab sheathing.

Board sheathing is fastened to the studding with enough nails to maintain panel integrity. Generally two nails per stud crossing are used, but fewer nails can be used if the form is used for only one pour. When forms are used for one pour, the lumber may be salvaged and used for other parts of the building.

Beam Side Installation

Beam sides are placed along the beam bottoms and tacked in place by nailing through the beam side into the beam bottom. They are held in place by a kicker. The kicker is usually made from a 2 by 4 or a rough 1 by 3. It is nailed to the shore heads with double-headed nails. Where the beam forms meet the columns or girder forms, they must be cut to length to fit the existing conditions. Care should be taken to fit the beam sides so that stripping can be easily accomplished. In some cases it may be desirable to fit the ends of the beam sides with a narrow wrecking strip that can be easily removed and replaced when the form is reused.

As the beam sides are put in place and secured, they are temporarily braced in a vertical position and ledgers are nailed to them. The ledger will support the slab joists and sheathing.

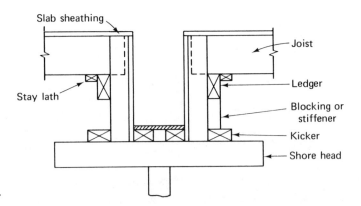

FIG. 5-7. Ledger Support.

Therefore, it must be securely fastened to the studs, and in some cases it may be necessary to provide extra support for the ledger by fitting blocks between it and the kicker (see Fig. 5-7).

SPANDREL BEAM FORMS

Spandrel beams require forms with two different sized beam sides. The inner form is low and runs to the bottom of the slab sheathing, but the outer form must be made higher to enclose the outer edge of the floor slab (see Fig. 5-8).

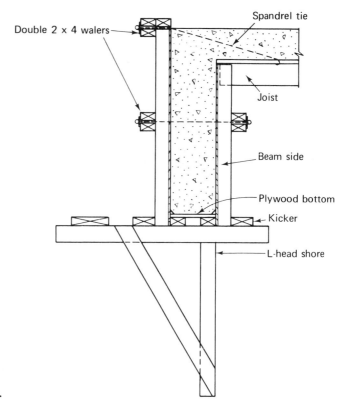

FIG. 5-8. Spandrel Beam Form.

Because they are located at the outside of the building, spandrel beam forms should be carefully aligned and braced. Their greater height may require waler and tie support. These are installed in the same manner as walers and ties for wall forms.

FORMS FOR FIREPROOFING STEEL BEAMS

If concrete is used to encase structural steel beams, the forms required may be suspended

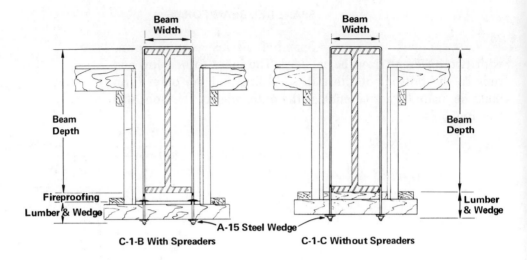

WEDGE HANGERS

FIG. 5-9. Beam Hangers. Courtesy Dayton Sure-Grip and Shore Co.

from the steel frame by means of wire hangers (see Fig. 5-9). These wire hangers are available with the same type of construction as snap ties and can be ordered to meet varying job conditions such as different beam sizes and different concrete form makeup.

Another type of hanger can be ordered for use with coil bolts. The coil bolt hanger can support heavier loads if each side is loaded equally. This type of hanger can be used on exposed concrete by providing a setback and using longer coil bolts (see Fig. 5-10).

If hangers are used to support the beam form, the beam bottom is made up in the same manner as for other beam forms, but it must have a parallel row of holes drilled at the proper spacing and proper center-to-center distance. When the beam bottom is hung in place, the hangers are placed through the holes and connecting hardware is temporarily installed to hold the beam bottom.

Beam bottom supports are placed beneath the beam bottom is hung in place, the hangers are placed through the soon as the beam bottom supports (usually 4 by 4's) are in place, the connecting hardware is installed.

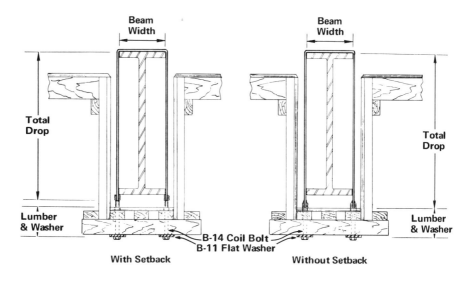

COIL HANGERS

FIG. 5-10. Beam Hangers. Courtesy Dayton Sure-Grip and Shore Co.

The first connectors that were installed before the bottom supports were in place can now be removed and replaced after the bottom supports are put in. Remaining bottom supports and hangers are connected before the beam sides can be installed.

The beam sides are placed on the bottom supports and tacked to the beam bottom. Kickers are nailed to the bottom supports to keep the beam sides against the beam bottom, and ledgers are installed as required in the same manner as for beam forms supported on shoring.

STRIPPING BEAM FORMS

All forms must be stripped carefully to prevent damage to the concrete surface. Wood wedges should be carefully driven between the form and concrete to release the form panels at one end. When one end is free, the form panel can usually be peeled off.

The form builder should make his forms with stripping in mind so that the forms will be built for safe, easy removal. As the panels or sections of the form are removed, they should be handed down and carefully stacked. Care should be taken to avoid damaging the contact face and edges of the panels.

Orders for stripping the forms will be given by the job engineer, who will determine when the concrete has reached sufficient strength to support the live and dead loads placed on it. If the job engineer does not determine stripping time on the basis of concrete strength, ACI Committee 347 suggests that under ordinary conditions the forms and supports remain in place for the minimum periods stated in Table 5-1.

Table 5-1. Minimum Time Forms and Shores Should Remain in Place

TYPE OF FORM	TIME
Arch centers	7-14 days
Beam and girder soffits	4-21 days
Floor slabs	3-10 days
Walls	12-24 hours
Columns	12-24 hours
Beam and girder sides	12-24 hours

The overall size of the form and the various functions it serves requires that the form be left in place for a greater or lesser period of time. For additional information on stripping and reshoring, refer to Appendix A.

Time, effort, and materials can be saved by training the form stripping crew in the order of form removal and stripping procedure. Care in handling panels cuts down on repair work and extends the life of the panels.

Reshoring should be done on one girder at a time as the forms are being removed. Reshores should be placed in line from floor to floor so that loads are transmitted in a straight line without causing stresses that the beams were not designed to bear. The location and spacing of reshores is governed by the plans and specifications. These should be followed carefully to avoid failures in the partially cured beams and floors.

REVIEW QUESTIONS

1. How do girder and beam forms compare?
2. List and define the various parts of a beam form.

3. How may beam forms be supported?
4. What precautions should be taken when placing mudsills?
5. How does the carpenter determine shore spacing?
6. How may shore length be adjusted?
7. How is the grade elevation of shoring checked?
8. What kind and amount of bracing is required on form shoring?
9. How may beam bottoms be installed?
10. How are beam sides installed?
11. How is the width of beam sides determined?
12. How do spandrel beam forms differ from interior beam forms?
13. How are beam forms supported when enclosing structural steel?
14. What precautions should be taken when stripping beam forms?
15. When is reshoring of concrete beams required?

FORMS FOR CONCRETE FLOOR SLABS
CHAPTER SIX

Forms for concrete floor and roof slabs are built on shoring that is constructed to the proper height and made strong enough to support the weight of workmen, reinforcing steel, equipment, concrete, and the formwork, with an adequate margin of safety. The manner in which the form is built is governed in part by the design of the floor slabs.

The floor slab form may be job built from lumber and plywood, or it may be made from a patented panel system. If the floor is to have joists formed in it, a combination of lumber, plywood, and pans will be needed to complete the formwork.

SHORING FOR FLOOR SLABS

Floor slab formwork is generally supported on a network of joists, stringers, and shores (see Fig. 6-1). The spacing of the supporting members is governed by the anticipated loading, the finish requirements, and the load-carrying capacity of the materials used. The forming plans should be referred to for determining the spacing of the shores, stringers, and joists, and all work should be done in accordance with shoring safety requirements (see Appendix B).

SHORING FOR FLOOR SLABS

FIG. 6-1. Floor Slab Shoring.

Single Post Shores

Single post shores may be made from 4 by 4's cut to length on the job and provided with wedges for final length adjustment (see Fig. 5-4).

Single post shores made of spliced 3 by 4's or spliced 4 by 4's can be adjusted with Ellis Clamps (see Fig. 6-2). Two Ellis Clamps are fastened to the lower half of the shore a minimum of 12 inches apart. These clamps are made with a solid rectangular collar and heavy malleable castings that are scored on the flat surface to grip the wood shore. To adjust the length of the shore, the upper portion is simply lifted either by hand or with the Ellis Jack, depending on the load. The clamps release to allow upward movement, but by keeping the clamp plates tight against the shore and removing the lifting force, the two halves of the shore are tightly clamped together by the downward force on the upper shore. A 16d nail driven into the shore just above the lower clamp ensures against slippage.

Shores are fastened to the stringers (purlins) with the Ellis purlin splicer or the Ellis slip in shore holder. Each is attached to the stringer with double-headed nails.

Shore with Clamps in Place

Ellis Jack

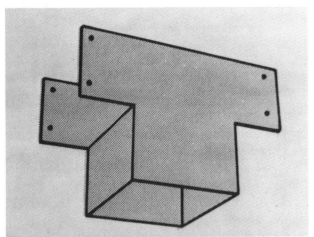
Ellis Clamp

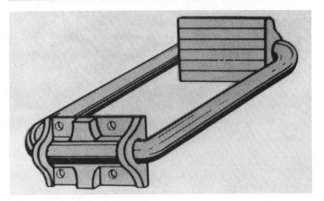
Ellis Purlin Splicer

FIG. 6-2. Job Built Single Post Shore.

SHORING FOR FLOOR SLABS 125

FIG. 6-2. Continued.
Ellis Slip-in Shore Holder in Place on Stringer

A number of companies manufacture adjustable single post shores made either of a combination of wood and steel or all steel. One type of adjustable shore made of wood and steel is illustrated in Fig. 5-3A. Shores of this type are easy to adjust in length and are equipped with a positive locking device. The

FIG. 6-3. Manufactured Single Post Shores. Courtesy Safway Steel Products.

built-in jack is convenient and does away with the need for carrying jacking equipment from shore to shore. Shores can be fastened to purlins by nailing through the top bearing plate, and if needed, shores can be fitted with a T-head using hardware manufactured for that purpose.

The all-steel single post shore in Fig. 6-3 is representative of shores of this type. It is manufactured in various lengths and load-carrying capacities. Each size has a flat steel base and cap with holes for attaching it to stringers, and all can be fitted with a U-head for attaching the shore to stringers or joists.

Steel shores of this type are adjusted to approximate length by placing a keeper pin through one of the holes in the upper tube of the shore. By rotating a screw adjusting collar, the shore can be adjusted to exact length.

Sills should be laid out for the shores as indicated on the shoring plan, and the shoring locations should be marked on the sills according to the spacing shown on the plan.

Before any shores are installed, they should be inspected for damaged parts. Wood shoring should be checked for splits, crushed grain, and rotted sections. Steel shoring should be checked for excessive rust, broken welds, and other damage. All defective shores should be set aside for repair or be discarded if the damage is extensive.

Shores should be adjusted to approximate length before they are erected. During erection, they should be set on the sill layout marks and nailed in place with double-headed nails. Diagonal bracing should be installed as the shores are erected. All shores must be plumb.

As shores are erected, stringers are placed on the shore head and fastened in place. All stringer ends should be supported on a shore or properly made ledger at the beam form. Following the completion of the vertical shoring and stringers, the joists and slab sheathing can be installed (see Formwork for Flat Slabs, pp. 132-147 and Formwork for Ribbed and Waffle Slabs, pp. 147-153.

Steel Frame Shoring

Steel frame shoring (see Fig. 6-4A), like all types of shoring, should be checked for damage and all faulty components should be repaired or replaced.

FIG. 6-4A. Steel Frame Shoring. Courtesy Safway Steel Products.

The legs of steel frames support considerable concentrated loads. Therefore, they must be placed on suitable sills (except when placed on concrete) in accordance with the shoring layout plans. Adjustment screws should be set to approximate length before setting up the shoring units.

The usual procedure for setting up steel frame shores is to locate and place the necessary sills and to mark the location of the frames. Next, the various adjusting screws, base plates, and braces are distributed along the marks indicating the tower location.

Following distribution of the various parts, the basic unit consisting of two frames, two sets of cross braces, and four leveling screws is assembled and set in place on the layout marks. The base plates should be nailed to the sills with double-headed nails; or, if 16d common nails are used, the nails should be bent over to hold the base plate and also facilitate removal. The basic unit should be leveled and plumbed so that additional frames placed on top of the basic unit will be level and plumb.

As tower erection proceeds, the various towers should be tied together with sufficient bracing to make the shoring rigid. Improperly built shoring is a major cause of form failure, but it can be avoided by following the "common sense" check list for final inspection of erected shoring equipment formulated by the Steel Scaffolding and Shoring Institute. This check list is given in the following paragraphs. Many of the recommendations can be applied to all types of shoring.

1. Check to see that there is a sound footing, or sill, under every leg of every frame on the job. Check also for possible washout due to rain.
2. Check to make certain that all base plates or adjustment screws are in firm contact with the footing or sill. All adjustment screws should be snug against the legs of the frame.
3. Obtain a copy of the shoring layout that was prepared for this specific job. Make sure that the spacings between towers and the cross brace spacing of the towers do not exceed the spacings shown on the layout. If any deviation is necessary because of field conditions consult with the engineer who prepared the layout for his approval of the actual field setup.
4. Frames should be checked for plumbness in both directions. The maximum allowable tolerance for a frame which is out of plumb is ($\frac{1}{8}$" in 3'). If the frames exceed this tolerance the base should be adjusted until the frames are within the tolerance.
5. If there is a gap between the lower end of one frame and the upper end of another frame it indicates that one adjustment screw must be adjusted to bring the frames in contact. If this does not help it indicates the frame is out of square and should be removed.
6. When two or more tiers of frames are used, each shall be braced to at least one adjacent frame.
7. While checking the cross braces also check the locking devices to assure that they are all in their closed position or that they are all tight.
8. Check the upper adjustment screw or shore head to assure that it is in full contact with the formwork. If it is not in contact it should be adjusted or shimmed until it is.
9. Check to see that the obvious mistakes of omitting joists, using the wrong size ledger or timber placed flat has not been made. Check the print to see that the lumber used is equal to that specified on the shoring layout. Check the general formwork scheme to make sure that it follows good standard practice for formwork.

10. If the shoring layout shows exterior bracing for lateral stability, check to see that this bracing is in place in the locations specified on the drawing. Check to make sure that the devices which attach this bracing to the equipment are securely fastened to the legs of the shoring equipment. If tubing clamps are used, make sure that they have been properly tightened. If devices for holding timber have been used, check to see that sufficient nails have been used to hold the bracing securely to the frame legs.

Flying Shores

A number of scaffold and shoring equipment companies have developed a system of flying shores for shoring the floors of multi-story buildings. The system utilizes various types of heavy-duty shores permanently fastened to a wood sill and adequately braced. The adjustable shore heads support a large beam or purlin, which in turn supports the joists and slab sheathing (see Fig. 6-4B).

FIG. 6-4B. Steel Frame Shoring. Courtesy Safway Steel Products.

The entire unit is handled and set by a crane. When set in place, the shore heads are adjusted to proper elevation and locked in place. When it is time to strip the slab forms, the shore heads are released and turned down 1 to 2 inches, and the form is rolled out to a location where it may be moved by crane to the next floor. This system saves time and costs.

Horizontal Shoring

The horizontal shoring system employs beams of adjustable length to replace closely spaced vertical shores (see Fig. 6-5). The horizontal shores can be supported on a ledger placed on vertical shores, beam sides, or other support. Use of the horizontal shore opens large areas below the formwork to workmen of various trades.

FIG. 6-5. Horizontal Shoring. Courtesy Safway Steel Products.

As with other types of shoring, the safety rules prepared by the Scaffolding and Shoring Institute should be followed. The horizontal shoring should be spaced in accordance with the shoring plans and supported on ledgers, stringers, or other supports that can safely carry the imposed loads.

With the use of special clips, horizontal shores can be used in place of joists to support plywood slab sheathing (see Fig. 6-6A). The clips prevent the plywood from moving during erection and pouring and are fastened to the plywood with only

SHORING FOR FLOOR SLABS 131

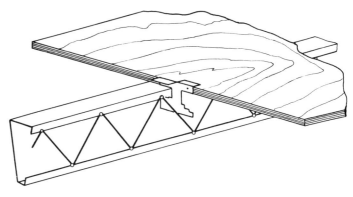

A. Span clips

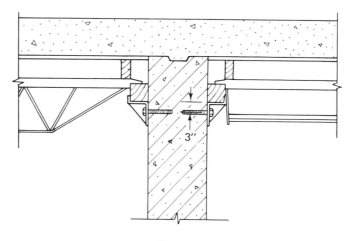

FIG. 6-6. Horizontal Shoring-Installation Details.

B. *Z* brackets

one nail. Two clips will normally hold an 8 foot joint, and one clip is usually sufficient to hold a 4 foot end joint.

If a concrete wall is used to support horizontal shores, steel brackets that support a 2 by 4 or 4 by 4 ledger can be fastened to the wall. The horizontal shore is supported on the ledger (see Fig. 6-6B). The steel bracket is fastened in place with $\frac{3}{4}$ inch coil bolts. The coil nuts must be accurately located in the wall to ensure proper alignment of all shoring members.

Concrete slabs are seldom of uniform thickness throughout the entire floor area. As with most shoring systems, horizontal shores may be adjusted to different slab thicknesses (see Fig. 6-7). In the area of drop pans around columns, an additional

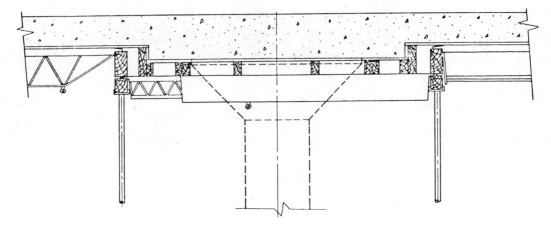

FIG. 6-7. Horizontal Shoring Supporting Dropheads.

ledger, which carries only the drop pan shores, is installed. This ledger is usually supported on vertical single post shores. The spacing of the vertical shores will be indicated on the shoring plan, which should be carefully followed to avoid costly form failures caused by omissions.

FORMWORK FOR FLAT SLABS

If horizontal shores are closely spaced, plywood slab sheathing is placed directly on the shores. If a network of vertical shores and stringers are used it becomes necessary to install closely spaced joists on the stringers in order to support the slab sheathing (see Fig. 6-8). The size and spacing of stringers, joists, and shores indicated on the forming plan should be closely followed. Any changes in forming dictated by job conditions should be approved by the job engineer to be sure the forms will perform satisfactorily.

Joists should be secured to the stringers, and measures should be taken to prevent joists from overturning. For joists, 4 by 4's are often preferable to 2 by 6's because they are less inclined to tip.

Slab sheathing should be nailed to the joists, using only enough nails to keep it in place. Usually one nail on each corner of a plywood sheet is sufficient. If boards are used for slab sheathing more nails are required, but as few as possible should be used to make it easier to strip the forms.

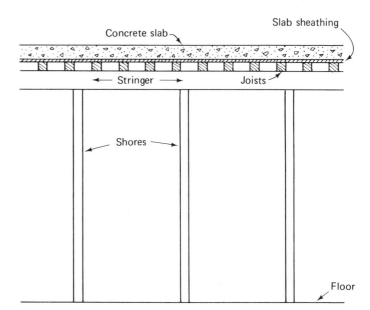

FIG. 6-8. Formwork for Flat Slabs.

Slab forms should not be stripped until the concrete has cured sufficiently. A qualified engineer should decide when to begin stripping the forms because a number of variables such as weather, temperature, size of structure, and loading must be considered (see Table 5-1).

Stripping slab forms prematurely in order to use them again can become costly if floor failure occurs. Stripping crews should be aware of this and avoid removing any forms until so ordered by the proper authority.

During stripping, the shore screws are released and the form is lowered from the concrete. In some cases the slab sheathing may stick to the concrete and can be left in place until the shores, stringers, and joists have been completely removed from an area. In many cases, however, the sheathing is lowered piece by piece as each section is released.

Slab Shore System

The Symons Slab Shore System utilizes standard steel ply form panels for "deck sheathing" and a system of stringers and adjustable shores to support them (see Fig. 6-9). The stringers are made with a movable ledger that allows the form panels to be removed while the shores and stringers remain in place to support the concrete slab during the curing stage. The system can be used to form square, rectangular, and round or irregularly shaped floors.

FIG. 6-9. Slab Shore System. Courtesy Symons Mfg. Co.

Erection Preparation. As the various components of the slab shore system are delivered to the job, they should be separated and stockpiled so that handling and extra movement is minimized during assembly and erection. Each component should be checked for serviceability and damaged parts should be set aside for repair or replacement.

Individual shores are made up from a base assembly, extensions when needed, and an adjustable shore head assembly (see Fig. 6-10). To determine the amount of fine adjustment needed on the shore head, the various components are assembled, and the shore head is turned on the threads until the desired overall length is attained. The threads between the inner and outer tubes of the shore head must not be visible after adjustment is completed. If threads become visible, the shore extension should be reset so that the shore head can be turned down to cover the screw threads.

After the proper setting for the shores has been determined, a jig may be attached to saw horses or a table and the shore heads may be rotated for fine adjusting to the proper length. The adjusted shore heads are attached to the shore and shore extension assembly with fast pins (Fig. 6-10) and set aside for further assembly.

FORMWORK FOR FLAT SLABS 135

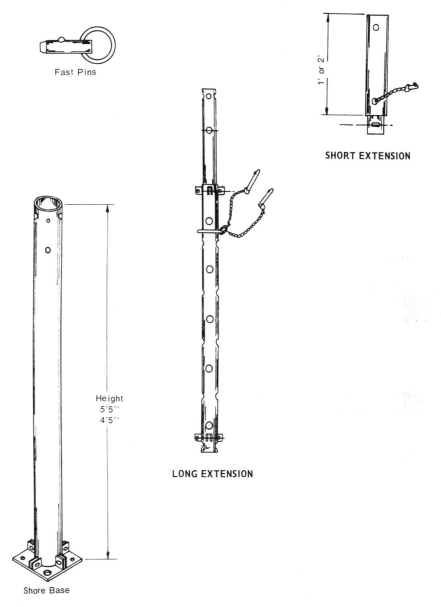

FIG. 6-10. Shore Components.
Courtesy Symons Mfg. Co.

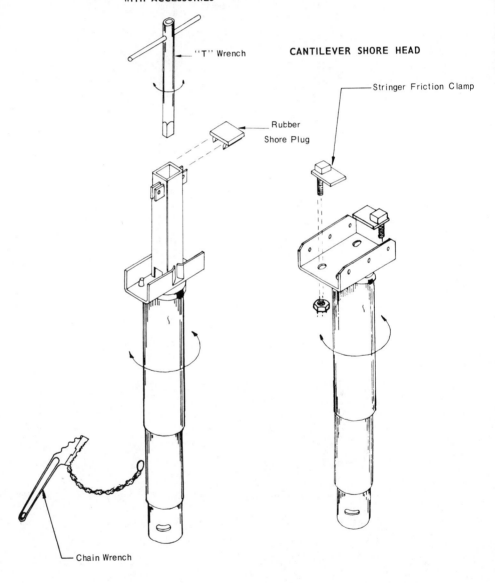

FIG. 6-10. Continued.

Stringers should be checked to be sure that the ledger angle is in the raised position (see Fig. 6-11) and that the fast pin is in place before attaching stringers to the adjusted shores.

In preparing to erect shores, two braces of the proper length are bolted together to form a pair. This procedure makes it easier to install the brace and helps maintain correct alignment.

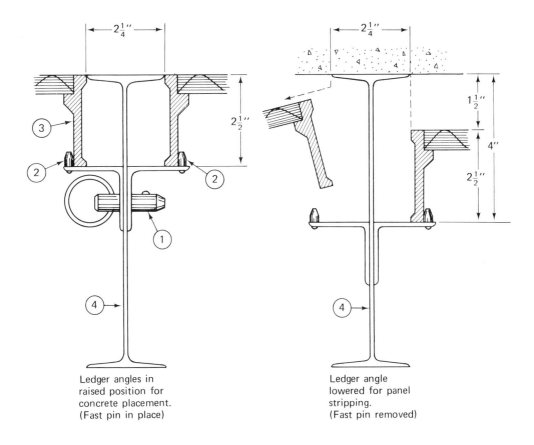

Ledger angles in raised position for concrete placement. (Fast pin in place)

Ledger angle lowered for panel stripping. (Fast pin removed)

(1) $\frac{1}{2}''$ Fast pin to be removed for stripping (one per stringer).
(2) Panel locking rivets secure panels on ledger angle.
(3) Panel section.
(4) Stringer (remains in place for slab support).

FIG. 6-11. Stringer Ledger Positions. Courtesy Symons Mfg. Co.

Shore Layout. With a measuring tape and chalkline, the center lines of the shores are laid out on the floor in accordance with the shoring plan. Along these layout lines, 2 by 6 sills are placed. The sills are needed to distribute shore loads over greater slab area, to provide shore stability, and to reduce the possibility of shore base shifting. Sills must be placed under all shores, but to save material and to provide access aisles, sills can be placed under pairs of braced shores (see Fig. 6-12).

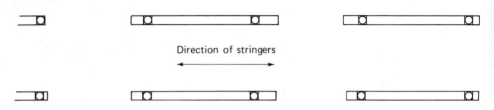

FIG. 6-12. Sill Plan. Courtesy Symons Mfg. Co.

Shore Erection. The initial tower consisting of four shores, two stringers, and four sets of proper length braces should be erected in a corner from which long runs can be made in two directions. Correct location and proper alignment of the first tower will simplify and speed the entire erection procedure. Therefore, extra care should be taken during its installation.

To build the initial tower, two identical bents, each consisting of two pre-adjusted shores, a stringer, and a pair of pre-bolted "S" braces, are assembled (see Fig. 6-13). The connection between the shore head and stringer is made positive by placing a 16d nail through the mating holes on the shore head and stringer. This nail can be removed after the bent is in a vertical position.

When assembling the bent, the stringers should be arranged so that the crescent slots are facing in the same direction to facilitate stripping. After assembly, the bents are erected on the sills and cross braces are installed between the bents. The shores can then be aligned with layout marks and nailed to the sills with double-headed 16d nails.

The shores should be plumbed in both directions and necessary adjustments made by moving the brace hardware. After the shores have been plumbed, the steel ply panels may be placed on the ledger angle (see Fig. 6-14). Panel side rails must

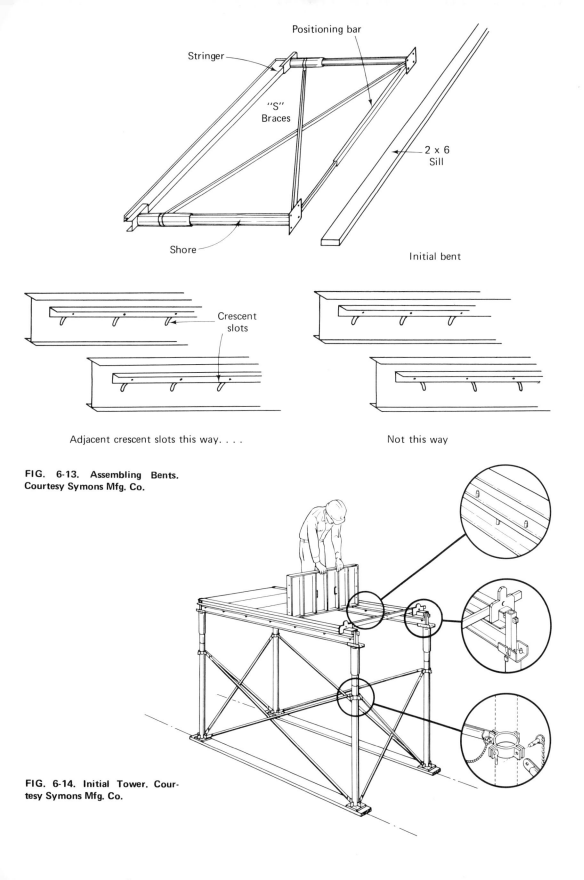

FIG. 6-13. Assembling Bents. Courtesy Symons Mfg. Co.

FIG. 6-14. Initial Tower. Courtesy Symons Mfg. Co.

FORMS FOR CONCRETE FLOOR SLABS 140

be placed at least 1 inch from the locking rivets on the ledger to prevent interference during form stripping (see Fig. 6-15A). As panels are placed, all 16d nails used for connecting stringers and shore heads should be removed.

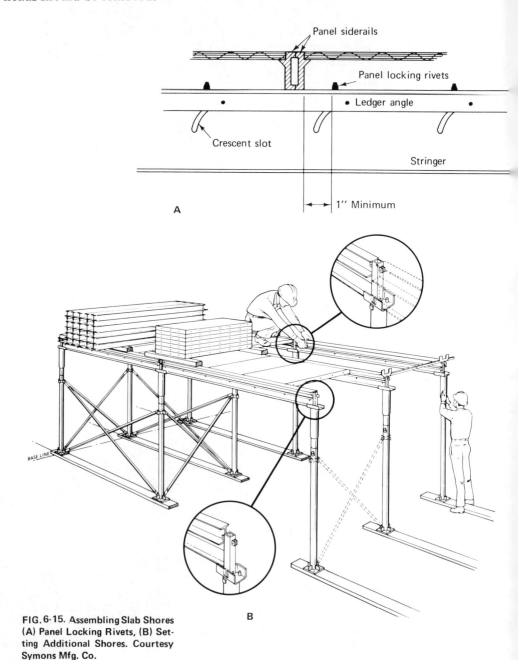

FIG. 6-15. Assembling Slab Shores (A) Panel Locking Rivets, (B) Setting Additional Shores. Courtesy Symons Mfg. Co.

After building the initial tower, subsequent stringers are attached to a single shore and the assembly is walked into position by one man on the floor while the man on the tower sets the stringer on the positioning pins of the shore in the completed tower (see Fig. 6-15B), and additional bracing is installed as required. If it is necessary to cantilever the stringers over the shore, a special cantilever shore head is used.

Final adjustment of the shore heads can be accomplished after the form is completed. This can be done by inserting a T-wrench in the end of the shore from the deck side and turning the shore head up or down as required. After the adjustment is completed, a rubber cap is placed in the top of the shore head to prevent grout leakage and to indicate that final adjustment has been made. The shore heads may be adjusted from below the deck with a chain wrench (see Fig. 6-16).

FIG. 6-16. Final Grade Adjustment. Courtesy Symons Mfg. Co.

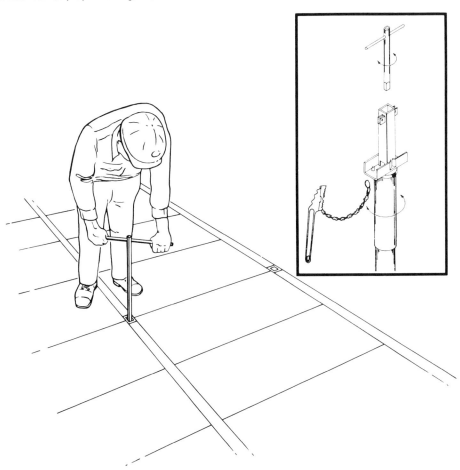

Form Stripping. A few days after the concrete has been placed, the cross braces may be removed from the shores and moved to a new location for reuse. A few cross braces may be required to remain in place in critical areas, such as where cantilevered shores are used. Removing the cross braces provides a larger unobstructed work area.

The form panels are stripped after the braces have been removed. To strip the panels, the fast pin that secures the ledger angle is removed and the ledger angle is lowered $1\frac{1}{2}$ inches with the special release wrench (see Fig. 6-17). With the angles lowered, the form panels can be raised over the locking rivets, shifted to the side, and lowered for piling and transfer to the next location.

Depending on ceiling height, a low scaffold may or may not be needed to facilitate panel removal. Many times a low

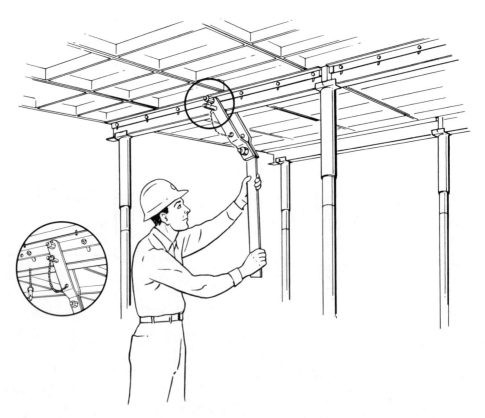

FIG. 6-17. Stripping Slab Shore Forms. Courtesy Symons Mfg. Co.

FORMWORK FOR FLAT SLABS   143

platform, as shown in Fig. 6-17, is of sufficient height and is easy to move as the form panels are stripped. If the ceiling height requires a higher scaffold, it may be advantageous to use a two-man crew with a rolling platform having locking casters. One carpenter remains on the floor to receive the panels as they are removed and moves the rolling platform as required, while the other carpenter remains on the platform to remove and hand the panels down.

The stringers and shores remain in place for 14 to 28 days, as required to allow the slab to cure. After the slab has cured, the shore heads are lowered about $\frac{1}{2}$ inch with a chain wrench, and the stringers and shores are removed for reuse. By using a special shore head, it is possible to remove the stringers and leave only the shores and selected stringers in place.

The Symons Manufacturing Co. maintains a staff of field representatives who provide help in solving forming problems

FIG. 6-17. Continued.

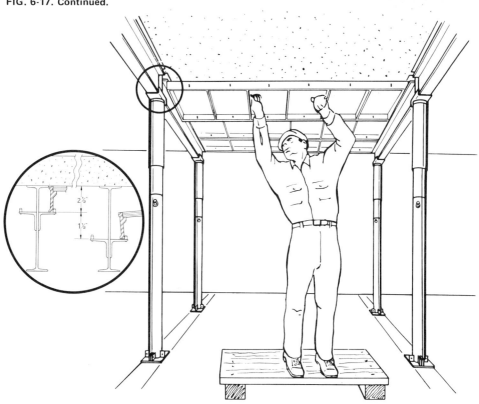

FORMS FOR CONCRETE FLOOR SLABS 144

and help train tradesmen in the use of Symons forms as required by unusual jobs or new forming techniques.

EFCO Slab Forming System

The Economy Forms slab forming system is designed and engineered to utilize EFCO all-steel form panels and the EFCO steel shore and stringer systems. In the EFCO system, the panels are supported on a stringer that can be lowered on the shore after the concrete has hardened. The panels and stringers can then be removed, and the shore is the only part of the form left in place after stripping.

To start the erection procedure, shores are located on the pre-established base lines, held in an upright position, and secured by horizontal struts. Temporary horizontal spacers are used to locate shores and to square the shore tower. After all the struts have been placed, the four shores are adjusted to proper elevation. Then, the diagonal braces are attached and secured in place (see Fig. 6-18).

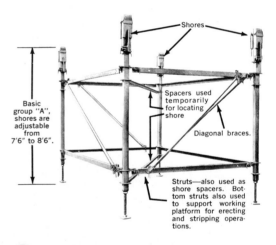

FIG. 6-18. EFCO Slab Forming. Courtesy Economy Forms Corp.

① Shores are placed in upright position and secured by struts. Shores are then adjusted to proper elevations, and diagonal braces put in place.

The erection of shores and struts can be continued following the basic procedure of locating shores with horizontal struts and temporary spacers. As the shores are set to the required elevation, the diagonal braces are added.

FORMWORK FOR FLAT SLABS 145

Trusses that support the slab and the form panels may be placed on top of the shores as the shore towers are completed (see Fig. 6-19). Care should be taken to see that all shore head locking pins are in place as the trusses are installed.

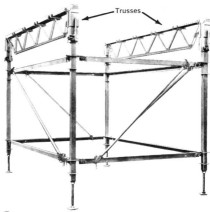

FIG. 6-19. EFCO Slab Forming. Courtesy Economy Forms Corp.

② Trusses are laid in place between top of shores and are held by gravity. No bolts or nails are required at any point.

After a sufficient number of trusses are installed, the form panels may be placed between the trusses (see Fig. 6-20). The panels for the first tower are set from below the trusses; but

FIG. 6-20. EFCO Slab Forming. Courtesy Economy Forms Corp.

③ Regular EFCO Forms are placed between trusses, to form the deck. All are held in position by gravity. The setup is now ready for concrete.

after panels have been placed on one or two towers, the panels can be stockpiled on the completed towers and placed from above.

The panels are held in place by gravity, but before concrete is placed the end panel of each tower must be secured to the truss with one plate clamp on each side, and each joint between panels must be secured by two plate clamps at the $\frac{1}{3}$ points.

Form Removal. After the concrete has hardened sufficiently, the form panels may be removed. To remove the form panels, the trusses and form panels are lowered by pulling the locking pins on the shore head (see Fig. 6-21).

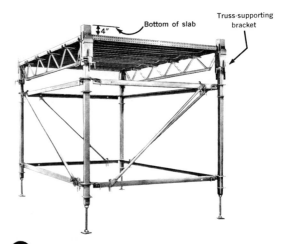

FIG. 6-21. EFCO Form Removal. Courtesy Economy Forms Corp.

④ When ready to strip the forms, the trusses and form panels are lowered 4 inches by pulling the pins in the truss-supporting brackets at the top of the shores.

With the shores lowered 4 inches, the plate clamps are removed and the form panels can be lifted off the trusses. As all panels are removed from each truss, the trusses may also be removed (see Fig. 6-22). The slab is now supported by the shores, and the struts and diagonal braces may be removed to make room for access aisles and working space for other trades.

Economy Form Corporation, like the Symons Company, maintains a staff of field representatives who provide help in

FORMWORK FOR RIBBED SLABS

After lowering, the form panels are removed, followed by the trusses. The shores need not be disturbed but can be left in place to support the concrete slab. The top of the shore acts as a form for the slab. **No reshoring is required.** To remove the shores, after struts and bracing have been removed, the jacks at base of shores are lowered about 1 inch.

FIG. 6-22. EFCO Form Removal. Courtesy Economy Forms Corp.

solving forming problems and help train tradesmen in the use of EFCO forms as required by unusual jobs or new forming techniques.

FORMWORK FOR RIBBED SLABS

Ribbed slabs are formed with metal, fiber, plastic, or fiberglass pans supported on top of some type of shoring arrangement. Under current safety regulations, the shoring system is covered with a continuous plywood deck. The location of the pan forms is marked out on the plywood, and the pan forms are nailed to the plywood. One typical forming system using flanged steel pan forms is illustrated in Fig. 6-23.

Installing Flange Forms

If the ribbed slab floor system is used, beam forms are built on top of shores in a conventional manner. Shores and stringers are erected between the beam forms in accordance

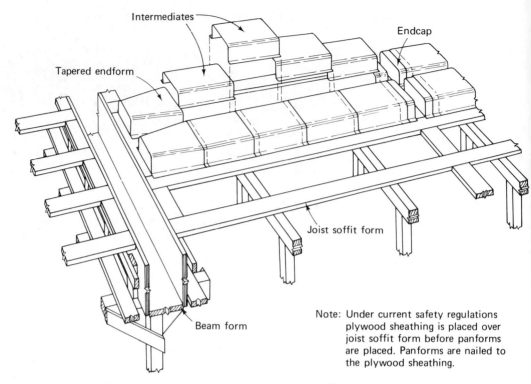

FIG. 6-23. Forming Joists with Flanged Pan Forms. Courtesy Ceco Corp.

Note: Under current safety regulations plywood sheathing is placed over joist soffit form before panforms are placed. Panforms are nailed to the plywood sheathing.

with the spacing shown on the forming plans. After the shores and stringers are set to grade and secured in place, the joist soffit forms may be located. The soffit form is usually made from nominal 2 inch thick lumber wide enough to form the bottom of the joist. These soffit forms become joists when covered with plywood.

As the plywood covers the soffit joists in a given area and is fastened in place, chalk lines are snapped to locate the edge of the pan flanges. The end pans in each row are set first and nailed in place along the chalk lines. Next, the intermediate forms are placed, starting from each end and working toward the center. The pans are allowed to overlap 1 inch to 5 inches and are nailed in place along the chalk line. Inserts for all piping that will pass through the floor are placed at this time.

As sections of joist forms are completed, they are oiled and work may begin on placing reinforcing steel and the various piping that will be embedded in the concrete. As with all slab forms, the grade elevation should be checked prior to placing concrete and the shores adjusted as necessary.

FORMWORK FOR RIBBED SLABS 149

Stripping Flange Forms

After the concrete has cured sufficiently, the job engineer or superintendent will give the signal to strip the forms. As the shores are released and removed, the stringers and joists that supported the soffit plywood sheathing are pulled free and piled aside for removal and reuse.

Generally, the soffit sheathing will stay in place as the stringers are removed because of the pans that are nailed to them and because of the slight bond between the concrete and the forms. Finally, the soffit forms and the pans are removed. Reasonable care should be exercised when removing the plywood soffit forms to avoid damaging the plywood and the flange forms that are fastened to the plywood. As pan forms are removed, they should be handed down for stacking, not dropped or thrown down. Mishandled forms become damaged and are difficult to reuse.

Long Joist Forms

Long joist forms avoid the need for shoring and placing stringers between the beam forms (see Fig. 6-24). They are

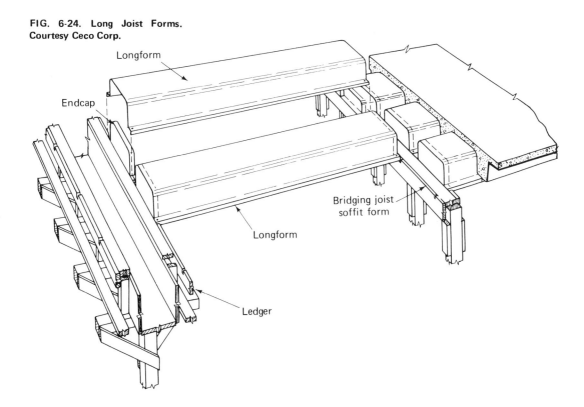

FIG. 6-24. Long Joist Forms.
Courtesy Ceco Corp.

supported by the beam form at one end and shoring at the other end. Steel long forms provide a smooth finish on the concrete joists, which are of uniform size from end to end. Fiberglass forms are available that provide the same smooth joist finish, and, in addition, end filler forms are available with tapered ends.

In installing long forms, a ledger must be fastened to the beam form to support the long forms and shoring must be erected to support the other end. This shoring also supports the bridging soffit form. The forms are oiled as placing progresses, and work on reinforcing steel and piping is started as soon as space permits.

The order for stripping long joist forms will be given by the job engineer or job superintendent when the concrete in the floor slab has cured sufficiently. The ledger supporting the joist forms is removed, and the long joist forms are pried away from the concrete at one end. Because of the size of these forms, two or more carpenters will work together in removing them and lowering them to the floor. Forms should never be dropped or thrown down because they will be needlessly damaged. As the forms are lowered to the floor, they should be stacked for cleaning and reuse. It is often advantageous to stack these forms on a four-wheeled cart so that they can be transported easily to the next placing area.

FORMWORK FOR WAFFLE SLABS

Standard steel dome pans for waffle slabs are available from a number of manufacturers. They are supplied on a rental basis, and many times the rental cost includes the forms, shoring, and stripping required for the job. These pans are either 24 inches or 36 inches square and form 19 inch or 30 inch square voids respectively. The depth of the pan will vary from 4 inches to 20 inches depending on the size of the form.

At least one manufacturer of fiberglass dome forms produces large domes measuring 48 inches, 60 inches, or 72 inches square. These pans form voids 40 inches to 63 inches square to a depth of 10 inches to 24 inches depending on the overall size of the pan.

FORMWORK FOR WAFFLE SLABS 151

Placing Waffle Forms

Waffle forms are supported by a network of shoring, stringers, soffit form joists, and slab sheathing (see Fig. 6-25). The installation of waffle forms requires that shoring and stringers be erected as discussed under Shoring for Floor Slabs, pp. 122-132, and Formwork for Flat Slabs, pp. 132-147.

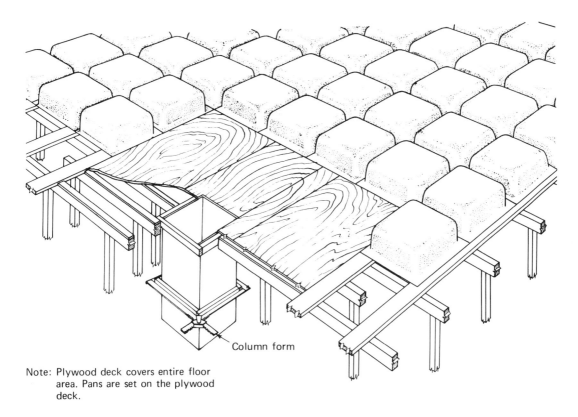

Note: Plywood deck covers entire floor area. Pans are set on the plywood deck.

FIG. 6-25. Forming Waffle Slabs.
Courtesy Ceco Corp.

Soffit joists are placed on top of the stringers and spaced on 24 inches, 36 inches, or other center distances depending on the size of pan used. After the soffit joists are fastened in place, plywood sheathing is placed over the entire area. The plywood is nailed to the joists to keep it in place. All end joints should occur on a joist. The plywood is fastened with as few nails as possible to make stripping easier. After a sufficient area is covered with plywood, chalk lines are snapped on the plywood in each direction to indicate the starting or base line. The chalk line represents the center line of the joist, and the edges of the pans are placed along this line. To hold the pans in place, a small nail is driven through holes at each corner of the pan into the plywood deck or soffit form.

Thick solid slabs are formed around the columns in all waffle slabs. The pan forms are left out in the area near the column, and the slab around the column is simply formed with the plywood sheathing placed over stringers and joists. Care should be taken to see that the stringers are supported on sufficient shoring and set at the proper elevation. The plywood in this area must be supported by a sufficient number of joists to avoid excessive deflection. If the forming plans are followed carefully, few difficulties should arise.

When placing plywood sheathing for the solid slab around the column, care should be taken to avoid placing a full plywood sheet with a cutout for the column at the center of the sheet (see Fig. 6-26A). Plywood placed in this manner is impossible to strip without damage. If forms are built with form removal in mind, a number of stripping difficulties can be avoided. Cutting the plywood panel as shown in Fig. 6-26B is a simple operation, and it makes the panel easy to strip from the hardened concrete (see Fig. 6-26C).

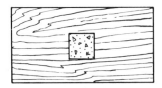
A. Improper cutout for column makes stripping impossible

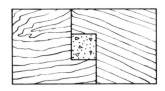
B. Panel in place around column

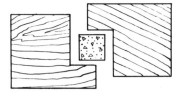
C. Cutting panels around columns in this manner makes removal easy

FIG. 6-26. Forming Around Columns.

After a reasonably large area is formed, the pans and slab sheathing is oiled, and the reinforcing steel, electrical, and other mechanical facilities that will be embedded in the concrete can be installed. Generally, the plumbing and electrical work will be completed before the reinforcing steel is placed, but on some jobs it may be desirable or necessary to perform mechanical and reinforcing work simultaneously.

FORMWORK FOR WAFFLE SLABS   153

Removing Pan Forms. After the concrete has cured and permission to strip the forms has been given, the shoring, stringers, joists and soffit sheathing are removed. Only the pans remain in place. The pans are held in place by the slight bond between the concrete and pan and by atmospheric pressure.

To remove the pans, air pressure is applied to a small hole at the top of the pan; properly oiled pans are easily released and removed (see Fig. 6-27). Two carpenters working together on a scaffold can remove pans systematically. As the pans are removed, they are handed down for stacking and reuse. They should never be dropped or thrown from the scaffold.

1. Air pressure is used to free the Steeldomes from the concrete voids.
2. Steeldomes are easily removed without damaging concrete surfaces.
3. Concrete surfaces are smoothly formed and require minimum finishing, if any.

FIG. 6-27. Removing Waffle Pans.
Courtesy Ceco Corp.

Pans that were not oiled properly or whose oil was washed away by rain before the concrete was placed will be difficult to remove. They will require much work with hardwood wedges, air pressure, and prying bars. The forms and the concrete will be damaged somewhat. The carpenter foreman or job superintendent can save a considerable amount of difficulty by checking the condition of the pans before the concrete is placed and calling for re-oiling if necessary.

REVIEW QUESTIONS

1. What are some of the various types of floor slab forms?
2. List and define the various parts of a floor slab form.
3. What are some of the safety precautions recommended for floor slab shoring?
4. What is a single post shore?
5. What is steel frame shoring?
6. What are flying shores?
7. What is horizontal shoring?
8. When is floor slab formwork stripped?
9. What are the various components of the slab shore system?
10. Outline the slab shore system layout and erection procedure.
11. Outline the slab shore stripping procedure.
12. Outline the procedure for building EFCO slab forms.
13. Outline the procedure for stripping EFCO slab forms.
14. How are ribbed slabs formed?
15. How are flange forms installed?
16. How are flange forms stripped?
17. How are long joist forms installed?
18. How are waffle slabs formed?
19. How are pan forms removed from the hardened concrete?
20. What precautions should be taken when forming slabs around columns?

FORMS FOR CONCRETE STAIRWAYS
CHAPTER SEVEN

Building formwork for concrete steps requires an understanding of stair layout and the functions of the stair form. This formwork ranges from simple, in which stairs are narrow and poured between two walls, to complex, in which stairways are wide and unsupported by walls.

The number of steps will also complicate the forming process. If there are few steps, the carpenter can stand on one floor and reach the other floor without scaffolding, and formwork used to support the stair slab need not be so heavy as the formwork needed to support longer stairways.

STAIR LAYOUT

One of the first steps in laying out forms for concrete stairs is to determine the unit rise and unit run. These may be obtained from the plans and specifications, but in many cases they must be determined by the form builder. If rise and run are given on the plan job conditions, they must be checked to determine if they conform to the plan. A typical stair section identifying the various stair terms used in the following discussion is illustrated in Fig. 7-1.

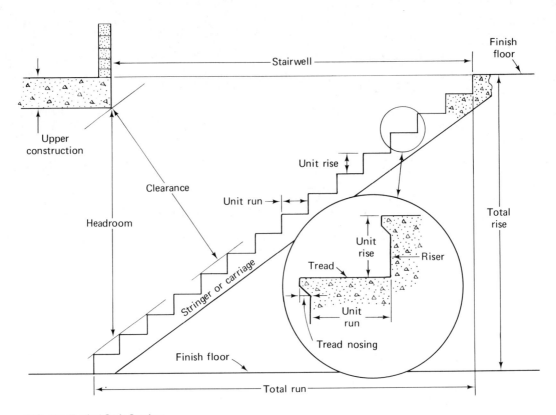

FIG. 7-1. Typical Stair Section.

To find the number of risers in a flight of stairs, the total rise is divided by 7 inches. The whole number obtained is divided into the total rise to determine the unit rise.

EXAMPLE

Total Rise $46\frac{1}{2}''$

$46\frac{1}{2} \div 7 = 6^+$ or 6 risers

$46\frac{1}{2} \div 6 = 7.75''$ unit rise

The unit run may be determined by applying a stair ratio formula if there are no other governing conditions. Two of the most widely used formulas are:

(1) unit rise + unit run = 17" to 18"
(2) 2 (unit rise) + unit run = 24" to 25"

These formulas can be rewritten, using averages, when determining unit run.

(1a) unit run = $17\frac{1}{2}"$ minus unit rise

(2a) unit run = $24\frac{1}{2}"$ minus twice the unit rise

Formulas 1 and 1a are the easiest to use and yield a less steep stair if the unit rise is over 7 inches. If the unit rise is less than 7 inches, formulas 2 and 2a will yield a less steep stair; and if the unit rise is over 7 inches, they will yield a steeper stair, but one that takes less horizontal space.

Using formula 1a, the unit run is determined in the following example

EXAMPLE

Total Rise $46\frac{1}{2}"$

Number of Risers 6

Unit Rise 7.75"

Unit Run = $17\frac{1}{2}"$ − unit rise

Unit Run = 17.50" − 7.75"

Unit Run = 9.75"

After determining the unit rise and unit run in the preceding manner, they may be used to mark the slope of the stair for the slab shoring, or they may be used to lay out the shape of the stair on stringers.

In some cases the total run of the stair will be governed by the location of previously placed landings. This is the case for the stairway in Fig. 7-2. The total run is 50 inches, and the total

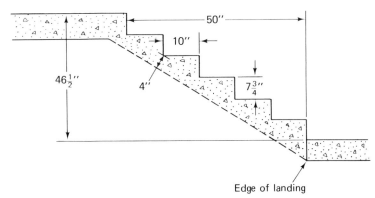

FIG. 7-2. Stair Section.

Edge of landing

rise is $46\frac{1}{2}$ inches. If six risers are used, the unit rise will be $7\frac{3}{4}$ inches and there will be five treads.

To get the unit run, the total run is divided by the number of treads. The result is checked with the stair ratio formula to be sure that a workable stair has been attained.

EXAMPLE

Total Rise $46\frac{1}{2}''$

Total Run $50''$

Number of Risers 6

Unit Rise $7\frac{3}{4}''$

Unit Run = $\dfrac{\text{Total Run}}{\text{Number of Treads}}$

Number of Treads 5

Unit Run = $50 \div 5$

Unit Run = $10''$

Check: Unit rise + Unit run = $17''$ to $18''$

$7\frac{3}{4}'' + 10'' = 17\frac{3}{4}''$

The results fall within the limitations of the stair ratio formula and can safely be considered a workable solution.

STAIRS ON GRADE

Stairs on grade are supported on sloping soil. They are usually outside stairs that are continuously exposed to the weather, but occasionally they may be interior stairs on the lower level of a building.

If concrete stairs are built on the ground, the upper and lower grades must be established before form construction can begin. These grade elevations may be established by previously placed floor slabs, or they may be established on stakes driven into the ground at a convenient location.

Before laying out the stair, the total rise must be checked. This can best be done with a level-transit and "story pole" (see Fig. 7-3A). If a transit is not available, the total rise can be

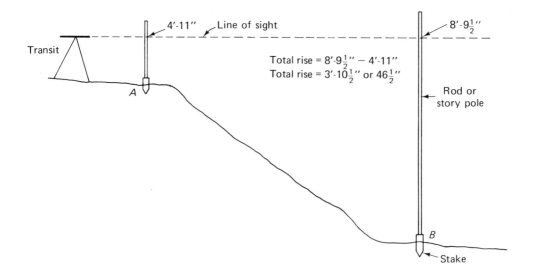

A. Using transit to determine total rise

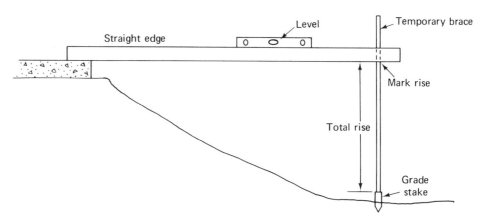

B. Using level and straight edge to determine total rise

FIG. 7-3. Finding Rise of Concrete Stair Placed on Grade.

checked by leveling a straightedge over the stair area and temporarily bracing it in place. The total rise can then be measured from the bottom of the straightedge to the lower floor or stake (see Fig. 7-3B). After the total rise has been determined, the number of risers, unit rise, and unit run are determined as outlined previously.

FORMS FOR CONCRETE STAIRWAYS   160

Stringer Layout

Stringer forms are usually made from 2 by 10 or 2 by 12 material. The shape of the stair is drawn on the stringer with the aid of the framing square and stair gauges (see Fig. 7-4). After the proper number of steps have been drawn, the stringer can be set in place and held with stakes and braces.

FIG. 7-4. Stringer Layout.

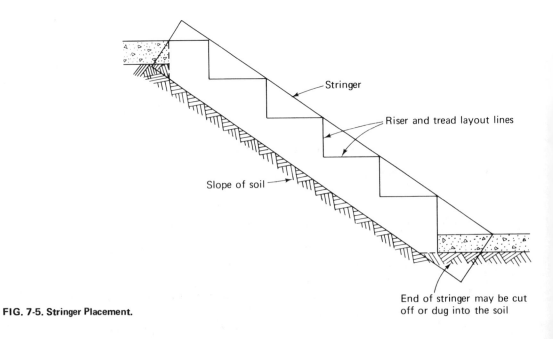

FIG. 7-5. Stringer Placement.

Care must be taken to locate the stringer accurately with the top of the upper riser in line with the upper slab or grade stake. The bottom of the lower riser should then be in line with the top of the lower slab or grade stake (see Fig. 7-5). If the top and bottom risers do not line up as required, the overall dimensions and calculations should be rechecked. If the total rise and total run vary from the original layout, the stringer should be re-marked using newly determined units or rise and run. However, if there is no change in overall dimensions or calculations, the stringer layout must be checked and the error corrected.

When the stringer is in its proper location, it is held in place with stakes and braces set on the outside of the stringer. To make form removal easier, nails should be driven through the stakes into the stringers.

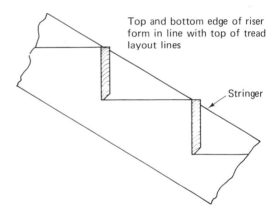

A. Risers nailed along layout lines

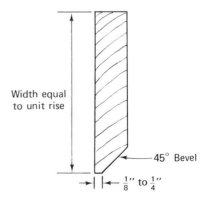

FIG. 7-6. Riser Installation.

B. Riser width

Riser Forms

Risers are formed by a riser board set in between the stringers and held in place on the layout lines with double-headed nails driven through the stringer into the ends of the riser form (see Fig. 7-6A). The riser form must be cut to exact length when installed in this manner, and it must also be ripped to the exact width (see Fig. 7-6B) because the stair form is filled with concrete to the top of the riser board.

To make concrete finishing easier, the lower edge of the riser forms should be ripped at a 45° angle. The clearance provided allows the cement finisher to finish the top of the tread all the way back to the riser face.

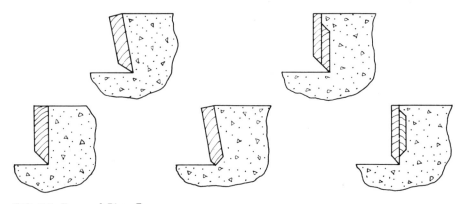

FIG. 7-7. Types of Riser Forms.

Risers for concrete stairs may take many different shapes. Some riser shapes with the riser forms in place are illustrated in Fig. 7-7. Notice that in each case the bottom edge of the riser form is beveled to allow the cement finisher to finish the tread up to the riser face, and that the top of the riser board provides a strike off for the surface of the tread. Risers are usually formed with nominal 2 inch lumber because it requires less bracing than 1 inch boards to resist the force of wet concrete; but riser forms also may be made from a combination of 1 inch and 2 inch stock.

Some designs require a non-skid edge on the treads. These non-skid materials may be set into the fresh concrete or provided for when the concrete is fresh to be installed later. If concrete stairways are faced with terrazzo, proper allowances must be made to allow for the thickness of the facing material.

As the width of the stair increases, it becomes necessary to brace the riser forms at the center to prevent the weight of fresh concrete from pushing the forms out of line and causing a bulge in the stair. This bracing can be accomplished by running a 2 by 4 at the center of the riser form from top to bottom of the stair and bracing the risers against it (see Fig. 7-12).

Form Removal

Riser forms are sometimes removed while the concrete is still green. This early removal makes troweling and finishing the riser face fairly easy. However, great care must be exercised when removing the riser forms to avoid damaging the concrete. Nails should be carefully removed from the center brace, and the nails driven in the ends of the riser forms should be carefully withdrawn. The riser forms can now be carefully pulled away from the concrete.

Green concrete is very weak and must be protected against damage. The edges of concrete steps are susceptible to damage during the early curing stages. Stairs may be barricaded, but many times barricades are by-passed and damage results. To avoid this damage, riser forms are often left in place until the concrete has cured sufficiently or until the danger of damage is past.

Leaving the riser forms in place for a longer period of time often makes them more difficult to strip, and it also makes finishing the risers more difficult. Therefore, some contractors prefer to strip the forms early and finish the risers as required. While this work is being done, the stairs are barricaded and no one is allowed to use them. The riser finishing is given 24 hours to cure, and the stairs are covered with riser and tread boards. These boards cover the full width of the treads and risers and are held in place by a 1 by 2 that runs the full length of the stair. This 1 by 2 is placed on both sides of the stair and is nailed to the nosing of each tread protector.

FORMING STAIRS BETWEEN WALLS

If stairs are formed between two walls the overall dimensions of the space allotted for the stair are checked, and the unit rise and unit run are determined as previously explained under Stair Layout.

After the unit rise and unit run have been determined, the shape of the stair may be drawn on the walls, and the slope of the underside of the stair slab is then established. Shoring is erected in the space below the stair soffit to the height required to support the stair slab. Because the shores are supporting a sloping slab, extra care must be taken when bracing them to prevent them from slipping out of place or collapsing while the concrete is being placed.

The risers for the stair are held in place by a stringer which is placed over the stair. In most cases this stringer will not be cut in the shape of the stair but will be a plank with cleats nailed to it to support the risers (see Fig. 7-8).

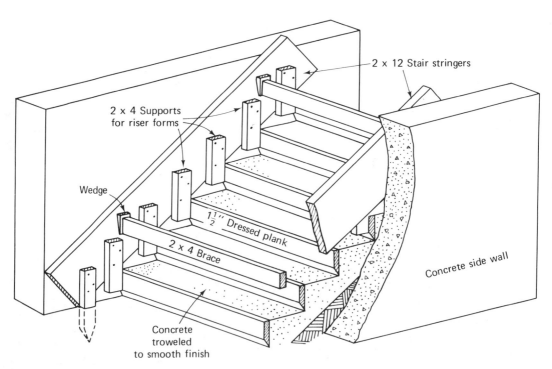

FIG. 7-8. Stair Form.

The procedure followed when building stair forms will vary somewhat to meet job conditions. However, the general forming procedure outlined here may be used for stairs between two existing walls if the top floor and landing are in place (see Fig. 7-9A).

FORMING STAIRS BETWEEN WALLS 165

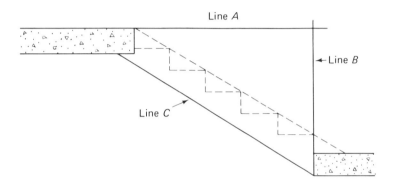

FIG. 7-9. Stair Section.

A. Plan section

B. Initial Layout

1. Check dimensions on plan with job conditions and make note of any variations.
2. Lay out line *A* on walls (Fig. 7-9B).
3. Mark total run on line *A*.
4. Draw a plumb line *B* at total run mark.
5. Check total rise at line *B* on both walls. If the total rise is not as shown on the plans, make the necessary adjustment in the unit rise so that all steps will be of equal height.
6. If desired, the risers and treads may be drawn on the wall. On long stairs, this may be done after the soffit form is in place.
7. Snap a line (*C*) locating the underside of the stair soffit.
8. Erect shores and joists to support soffit sheathing. The

FORMS FOR CONCRETE STAIRWAYS 166

type of shoring used may vary from 4 by 4's supported on wedges to commercial type shores with jack screws, and the size of joists will vary with the spacing of the shores and the thickness of the concrete slab.

9. Install soffit sheathing. When placing soffit sheathing, use as few nails as possible, and remember that this material must be removed from beneath. Install sheathing in a manner that will make removal easy.
10. After the slab sheathing is in place, the reinforcing bars are installed and before the riser forms are installed. The iron workers should be alerted in advance so that they will be ready to place the steel as soon as the slab form is completed.
11. Install supports for risers. These supports may be cleats nailed to the wall at each riser mark. To avoid puncturing the wall with many nail holes, a 2 by 6 or 2 by 8 plank may be fastened to the wall along the slope of the stair and riser support cleats nailed to the plank (see Fig. 7-8).
12. Rip risers to proper width. The top edge of the risers should be straight. The bottom edge should be beveled to permit finishing of treads.
13. Cut risers to fit between walls. To aid in removing riser forms, the back side of the risers can be cut away (see Fig. 7-10).

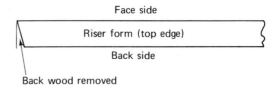

FIG. 7-10. End Cut on Riser Form.

14. Install risers on cleats by toe nailing from side away from concrete. This is necessary to make form removal easy and to avoid having nails in contact with the concrete, which will blemish the concrete surfaces.
15. Brace risers with 2 by 4 running down the center of the stair (see Fig. 7-12).
16. Check formwork.

Before concrete is placed, the stair form should be thoroughly checked to be sure that all dimensions are correct and that the entire structure is sufficiently braced.

FORMING STAIRS BETWEEN WALLS 167

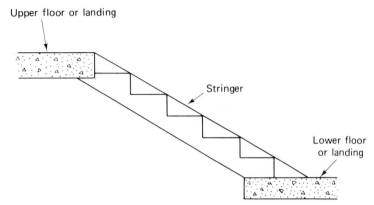

FIG. 7-11. Stringer in Place.

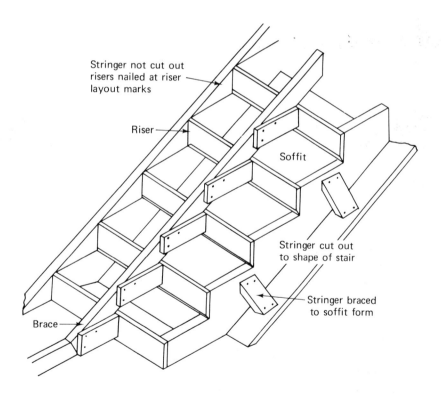

FIG. 7-12. Bracing Risers and Stringers.

FORMING STAIRS WITHOUT SIDEWALLS

Building a stair without supporting sidewalls complicates the layout problem slightly because there is no surface on which to lay out the shape of the stair. The location of the top and bottom risers must be determined and the shoring erected between these points.

The general procedure outlined here, which may vary with job conditions, can be followed when building forms for most stairs without supporting sidewalls.

1. Check dimensions on plan.
2. Check total rise between floors.
3. Check total run.
4. Check both floors for level.
5. Adjust unit rise and unit run if necessary.
6. Lay out stringers. Stringers may be laid out on 2 by 10, 2 by 12, or other stock. The width of the material used will vary with the thickness of the stair slab and the slope of the stair. Filler strips may be added to the bottom of the stringers to add thickness to the slab and avoid using excessively wide material.
7. Cut upper end of stringers to fit against the top floor.
8. Cut the lower end of the stringer to fit the lower floor. The stringer can be cut to the shape of the stair, but in most cases it is not cut in order to save labor and material. When the stringer is cut out, the risers are nailed to the riser cuts, but if the stringer is not cut out the risers are nailed between the stringers along the layout lines.
9. Put stringers in place against top floor and secure in place (see Fig. 7-11).
10. Erect shoring and joists to fit the underside of the stringers. Make allowance for sheathing to fit between the top of the joists and bottom of the stringer.
11. Reinforcing steel can now be installed as outlined on p. 166.
12. Rip risers to proper width and cut them to the required length.
13. Install risers along layout marks.

14. Brace risers and stringers as required (see Fig. 7-12).
15. Check formwork for dimensional accuracy and necessary bracing before placing concrete.

REVIEW QUESTIONS

1. What is one of the first steps in laying out a stair form?
2. Define the various stair terms in Fig. 7-1.
3. How may the number of stair risers be determined?
4. How is the stair ratio formula used?
5. Find the number of risers, unit rise, and unit run for a stair with a total rise of $139\frac{1}{2}$ inches.
6. Find the number of risers, unit rise, and unit run for a stair with a total rise of 135 inches and a total run of $165\frac{3}{4}$ inches.
7. Outline the stringer layout procedure.
8. Describe the layout of riser forms.
9. When are stair forms removed?
10. How are stairs protected after forms are removed?
11. Outline the procedure for forming stairs on grade.
12. Outline the procedure for forming stairs between walls.
13. Outline the procedure for forming stairs without sidewalls.
14. Why must all stair forms be carefully braced?

STOREFRONT CONSTRUCTION AND FINISHING
CHAPTER EIGHT

Each storefront presents a unique problem to the carpenter and other tradesmen because each is different. Plans and details must be carefully studied before work begins, and the builders of storefronts soon realize that they will apply a good portion of their knowledge of materials and construction details before the job is completed.

Storefronts often require the work of a number of different tradesmen. Many times tradesmen from a number of trades may have the expertise to do a certain phase of storefront work. As a result, jurisdictional disputes may break out and delay the job. Because jurisdictions change with local customs, it is sometimes difficult to determine which trade should perform certain phases of work. The storefront work discussed in this chapter is work normally done by carpenters.

BUILDING A STOREFRONT

Most commercial buildings are designed in a way to allow a wide variety of storefront design and construction. The roof or ceiling above the storefront is usually supported by beams and columns that leave the entire front area open. This area is "filled" in by a variety of materials to give the front its unique appearance.

Storefront Plans

Before attempting work on a storefront, a thorough study of the plans should be made. As work progresses on a building, various design details are changed. Therefore, the carpenter doing storefront work should be sure he has a copy of the latest set of plans and that all corrections and last minute changes are included on the plans. If the plans are not checked for revisions before actual construction is started, it may become necessary to dismantle and rebuild parts of the storefront before the entire job is completed.

Plate Layout. Stud walls in storefronts may be framed from wood or metal studs or a combination of both. Masonry veneer, if used, is installed after the stud wall is in place. Building a wood frame wall is discussed in the following paragraphs, and the layout for walls built with metal studs is similar (see Metal Stud Partitions, pp. 198-200).

After studying the plans, the storefront walls can be laid out in accordance with the plans. The walls are positioned on the floor and chalk lines are snapped between locating marks to show the outline of the wall. Plates may be fastened along these lines, and door and window openings and stud locations are marked on the plates (see Fig. 8-1). As the bottom plate is being marked, a similar top plate should be made.

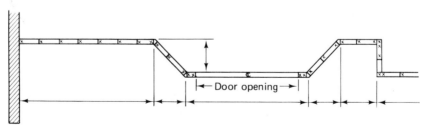

FIG. 8-1. Wall Layout. Wall plates located according to plan dimensions

Story Pole Layout. A straight 2 by 4 long enough to contain the full height of the wall is selected and marked with the various window heights, door headers, and other features (see

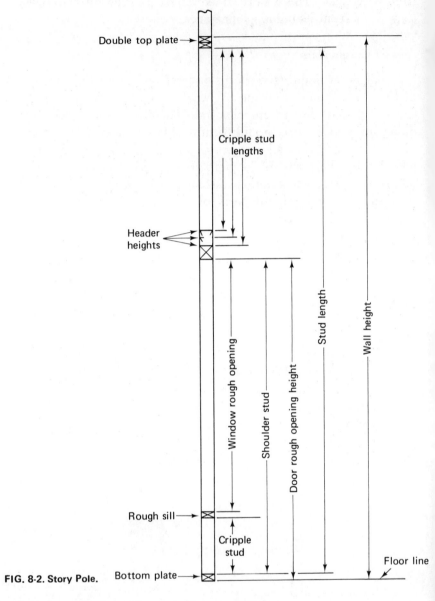

FIG. 8-2. Story Pole.

Fig. 8-2). The story pole is marked out in accordance with the building plans to obtain the exact length of regular studs, cripple studs, and other parts of the wall.

To lay out a story pole, the plans must first be studied and the rough openings and header sizes for the doors and windows determined. The floor elevation is marked near one end of the story pole and the bottom plate, wall height, top plates, ceiling height, door and window headers, rough sills, and other features are marked in. The completed story pole is used to determine the length of various studs and provides a permanent full-size record of the various wall features.

Fabricating the Wall

As the wall layout is completed, the various studs and headers are cut from stock length material. The cut-to-length pieces are distributed along the length of the wall and nailed in place in accordance with the layout marks.

Whenever possible, the wall will be fabricated by face nailing the top and bottom plates to the studs. The stud wall assembly with door and window rough framing may be squared and have sheathing applied before it is raised and fastened in place.

Job conditions such as columns and piping often do not allow the storefront wall to be fabricated and raised in place. It may then be necessary to fasten top and bottom plates in place and toenail the studs to the plates. This procedure requires more time than face nailing and is avoided whenever possible, but if it is impossible to raise full wall sections, there is no alternative.

Window Display Areas. Window display areas are made up of short supporting walls, floor joists, and flooring. The short supporting walls are usually made with single top and bottom plates with 2 by 4 studs placed 16 inches O.C. The joists in the display area also are usually placed 16 inches O.C. Their size will vary with the span, but in most cases 2 by 4 or 2 by 6 material is of sufficient size.

The flooring in the display area is usually either $\frac{1}{2}$ inch or $\frac{3}{4}$ inch thick plywood. When the display area is finished, the plywood is usually covered with a resilient tile or carpeting, but occasionally the floor is simply painted or coated with some other wood finish.

Installing Wood Frames. Wood door and window frames are installed in the rough openings and fastened to the rough framework by nailing through the exterior casing of the frame.

The door frame is set in the opening, and the head jamb is checked for level before any nailing is done. If the head jamb is not level, the lower end of the side jamb on the high side must be shortened enough to make the head jamb level.

After the head jamb is leveled, it is fastened to the wall. The side jambs are plumbed and straightened with a straightedge (see Fig. 8-3) and fastened to the wall in the same manner as the head jamb.

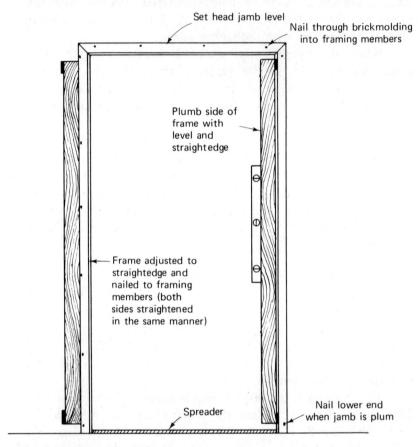

FIG. 8-3. Door Frame Installation.

Window frames are placed in the rough opening with the sill or head jamb held to the proper elevation as required by job conditions. The head jamb is leveled and fastened to the wall framework. On large window frames, a straightedge is needed to level the frame. Side jambs are plumbed and straightened in the same manner as are door frames.

After the frames are installed, the sills are blocked to provide extra support, and the side jambs are shimmed and nailed to the wall framework. This procedure is very important with frames for large windows to ensure that they can carry the weight of the window unit without excessive sagging.

Exterior Finish

Storefront exteriors may be finished in a wide variety of different materials. Some of those commonly installed by carpenters include wood siding, plywood paneling, prefinished panels of various types, and enameled steel panels.

Wood Siding. Wood siding is installed on storefronts in the same manner as on residential jobs. Its purpose is to weatherproof the structural parts of the wall and to give the wall an attractive finished appearance. In some areas building codes may require that a $\frac{5}{8}$ inch thick gypsum board sheathing be placed directly below the siding as a fire barrier. As building codes change among various areas, acceptable installation procedures will change.

If wood siding is used, it is usually of cedar or redwood and may be surfaced smooth or rough textured. Various patterns of siding are available and often more than one pattern is used on a store front to provide a unique appearance.

FIG. 8-4. Nails for Siding. Courtesy California Redwood Assoc.

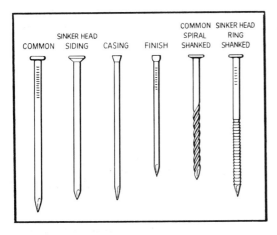

Siding installed on the exterior of a building should be fastened with high-strength aluminum nails, hot-dipped galvanized nails, or nails made of stainless steel. These nails should be long enough to penetrate through the siding and a minimum of $1\frac{1}{2}$ inches into the wood framework. If $1\frac{1}{2}$ inches of penetration cannot be attained, ring shanked nails should be used. The various kinds of nails used for siding application are illustrated in Fig. 8-4.

Bevel siding should be applied so that the exposure of the various siding courses is approximately equal. The siding courses should have a minimum overlap of 1 inch, and the nails should be driven just above the thin edge of the underlying siding to permit movement caused by changes in moisture content of the siding due to weather conditions. Rabbeted bevel siding should be nailed about 1 inch above the thick edge, and about $\frac{1}{8}$ inch clearance should be left between siding courses to allow for changes in the siding due to changes in moisture content (see Fig. 8-5).

Tongue and groove siding materials 4 inches, 5 inches, or 6 inches wide may be blind nailed or face nailed. Blind nailing eliminates the need for setting the nails and filling the holes, results in a nail free finish surface, and is generally preferable to face nailing. Standard 6d finish nails may be used on interior work, but for exterior work hot-dipped galvanized or other rust-proof finish nails must be used. In blind nailing, only one nail per bearing driven through the tongue is necessary (see Fig. 8-6).

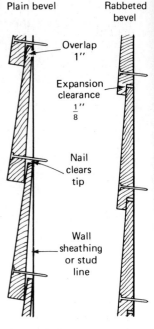

FIG. 8-5. Nailing Bevel Siding.

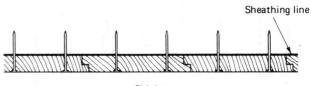

FIG. 8-6. Nailing Tongue and Groove Siding.

Note: These sidings may be installed either horizontally or vertically

BUILDING A STOREFRONT 177

Tongue and groove sidings 8 inches and wider should be face nailed with two nails per bearing, but narrower boards may be face nailed with only one nail per bearing. On interior work, the nails should be set below the wood surface to allow for putty or wood filler. On exterior work, the nails may be set and holes filled, or the siding nails may be driven flush with the surface and left to be painted with the siding.

Board and batten siding is often made from rough-sawn cedar, redwood, or other lumber to present a rustic appearance. These boards should be surfaced on one side (S1S) so that they

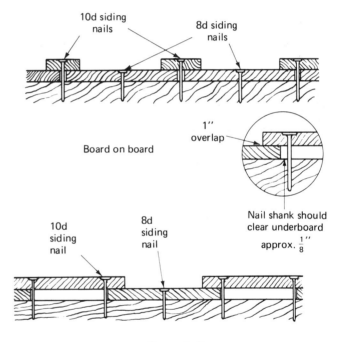

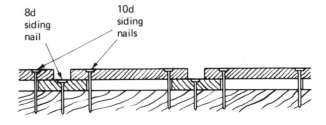

FIG. 8-7. Board and Batten Sidings.

are of uniform thickness. The surfaced side is placed against the wall and the rough side is left exposed.

The underboards are nailed to the wall with one 8d siding nail per bearing driven through the center of the board, and the boards are spaced approximately $\frac{1}{2}$ inch apart. Nailing in this manner allows the boards to swell and shrink without cupping or splitting as the moisture content changes. Batten strips are fastened over the space with one 10d siding nail per bearing driven between the underboards (see Fig. 8-7).

If the board-on-board batten system is used, the underboards and overboards are approximately the same width. The underboards are nailed with one 8d nail per bearing. Overboards should overlap the underboards by at least 1 inch and are nailed with two 10d nails per bearing. The nails should clear the underboards by about $\frac{1}{4}$ inch to allow for expansion due to moisture changes (see Fig. 8-7).

Channel pattern siding, which is a shiplap pattern, should be face nailed. Siding boards up to 6 inches wide may be fastened with one nail per bearing. Boards 8 inches wide and wider should be fastened with two nails per bearing. Nails should be placed about 1 inch from the edges of the boards. The overlapping joints should not be nailed together because the joint provides the necessary allowance for expansion and contraction (see Fig. 8-8).

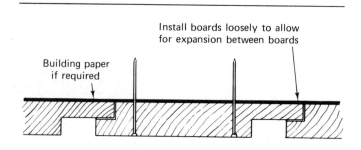

FIG. 8-8. Channel Siding.

Boards 8" and wider require 2 nails per bearing
Boards 6" wide require only 1 nail per bearing

Plywood Siding. Many patterns of plywood siding are available in 4 foot by 8 foot sheets and some patterns are available in longer lengths. These are generally fastened by face nailing. Care must be taken to apply plywood sidings in a manner that will make all joints weatherproof.

Vertical joints on plywood siding may have a shiplap joint, which is built into the panel, but vertical joints may be simple butt joints or butt joints covered with a batten strip (see Fig. 8-9).

Horizontal joints on plywood siding may be lapped, flashed, or shiplapped to make them waterproof. The lapped joint should be made in a way that will allow the top board to overlap the lower board by 1 to 2 inches.

Horizontal butt joints must be flashed to prevent water from running behind the lower panel. The flashing should be made in a manner that will allow it to fit neatly on the face of the panel and project behind the upper panel a minimum of $\frac{3}{4}$ inch (see Fig. 8-9).

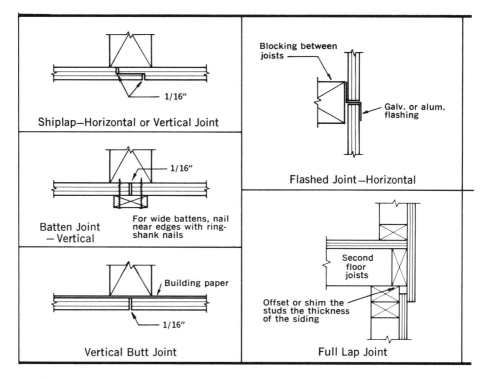

FIG. 8-9. Plywood Siding Joint Details. Courtesy American Plywood Association.

Colored Panels. Prefinished panels of asbestos-cement with a mineral color fused to the surface and of porcelain enameled steel are often installed by carpenters. The procedure followed when installing these panels will vary with the material and the type of installation. Details on the shop drawings should be followed carefully. If the panels are held in place by moldings, the usual procedure is to prepare the wall with furring strips as required and to apply the vertical and horizontal moldings in preparation for panel installation.

If panels will be held in place with an adhesive, the wall surface must be prepared by removing all grime, dust, and loose particles. The surface must be sound and recommendations of the adhesive manufacturer should be followed carefully.

If the panels will be placed over furring strips, a furring strip must be installed behind each joint to provide solid backing for the joint molding. Furring must be shimmed as required to cause all furring on the wall to lie in one plane.

After the furring has been installed, the usual procedure is to install the applicable vertical moldings (see Fig. 8-10A). Second, the lowest horizontal molding is installed. Third, the next to lowest horizontal molding is set into place slightly above its final location. This molding is carefully tapped into place after the panel is in place, leaving enough room so that the panel can expand.

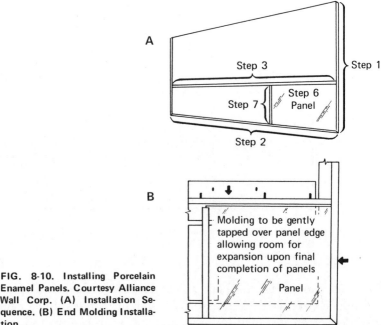

FIG. 8-10. Installing Porcelain Enamel Panels. Courtesy Alliance Wall Corp. (A) Installation Sequence. (B) End Molding Installation.

When the necessary moldings are in place, a bead of caulking is run in the channels (step 4). The next step is to apply the adhesive as required (step 5). Following the application of adhesive, the cut panels are set in place in the molding channel, with a $\frac{1}{16}$ inch space for expansion being left between the panel and the web of the molding (step 6). Finally (step 7), the end molding is cut to size, the channel is given a light bead of caulking, and the molding is set in place (see Fig. 8-10B). After the first panel is in place, the applicable steps are repeated until the installation is completed.

Various moldings used for procelain panel installation are illustrated in Fig. 8-11. Notice how the channel in each type covers the edge of the panel and allows room for caulking and panel expansion. Caulking is required to make the joints waterproof, and the expansion clearance is needed to prevent the panels from buckling or bulging out when heated by the sun.

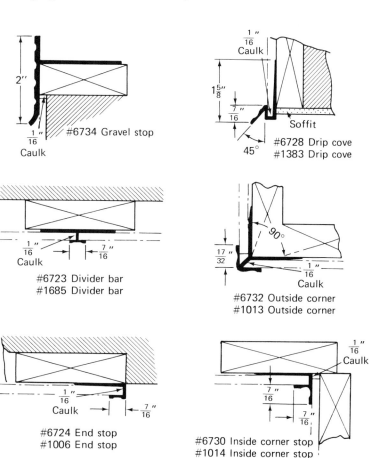

FIG. 8-11. Moldings for Porcelain Enamel Panels. Courtesy Alliance Wall Corp.

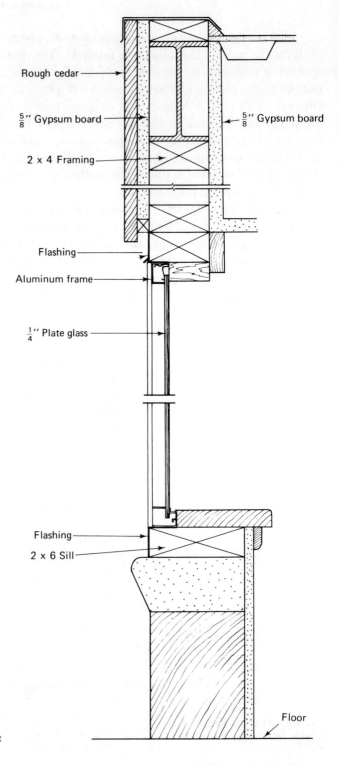

FIG. 8-12. Typical Storefront Section.

PREPARING STOREFRONT FOR METAL WINDOW SASH

Many storefronts are made almost entirely of glass set in metal frames. These metal units and glass are installed by other trades, but the carpenter usually prepares the opening for the metal frames.

Typical storefronts may require a cripple (short) stud wall or masonry below the window opening and some type of short wall framing above the opening (see Fig. 8-12).

The storefront opening requires a wood sill that is fastened to the masonry sill with anchor bolts. The area above the window must be filled in with 2 by 4 framing so that the rough opening is 8 feet 6 inches high. Cripple studs are cut to the proper length, and a 2 by 4 plate is nailed to each end of the studs. The assembled cripple wall is fastened to the steel beam by powder-actuated fasteners or other suitable means. If it is possible to pass wires over the top of the steel beam, the wall assembly may be held in place by tie wires.

After the wall assembly is fastened in place, it must be braced to straighten it and keep it aligned. This can be accomplished by stretching a line from end to end along the bottom edge of the wall. Braces of 1 by 4 or 2 by 4 lumber are placed at 4 foot to 8 foot intervals and run back to some solid framework.

When the wall has been straightened, gypsum board is applied over both sides for fireproofing and the finish material may be applied to the outside so that the exterior may be made weatherproof.

The metal work and glass are installed by other trades. After that work and other interior work is completed, the carpenter will install the interior trim.

INSTALLING DOOR BUCKS

A door buck is a door frame installed in a masonry wall. This frame may be made of steel, finished lumber, or framing lumber. Frames made from 2 inch thick framing lumber are often called rough bucks.

Door bucks are installed in accordance with dimensions given on the plan. The usual procedure is to snap chalk lines on the floor to show the outline of the wall and to locate the door openings between these chalk lines. If steel bucks are used, they are set in place in accordance with the layout marks and held in place by fastening the sill anchor bracket to the floor (see Fig. 8-13). The frame is braced with at least two diagonal braces to hold it plumb. Spreaders are placed between the side jambs to keep them from being pushed together by the concrete block (see Fig. 8-14).

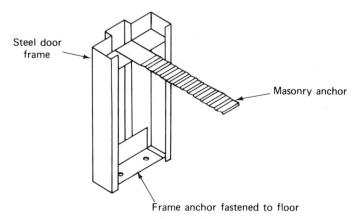

FIG. 8-13. Fastening Steel Frame to Floor.

FIG. 8-14. Bracing Steel Door Frames.

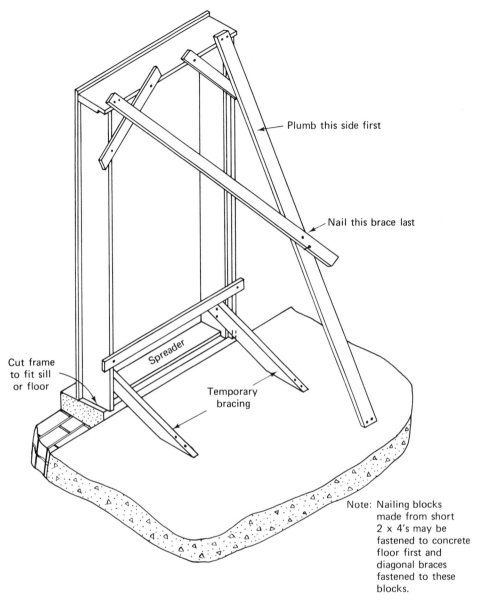

FIG. 8-15. Bracing Wood Door Frames.

Wood bucks are installed in much the same manner as steel frames. Steel angles fastened to the back of the wood buck may be fastened to the floor when the buck is in place; or if steel angles are not used, the wood spreader may be fastened to the floor to hold the buck in its proper location. Spreaders and braces are installed in the same manner as for steel bucks (see Fig. 8-15).

If finished wood frames are used, the installation procedure is similar to that for wood bucks, but extra care must be taken because of the finished nature of the material. These wood frames must be protected from damage. If this is impractical, rough bucks should be used and the wood frames installed after the danger of excessive damage no longer is present.

Arch Centers

Arch centers are temporary supports built of wood or other material to provide a form on which a masonry arch is built. Arch centers may be made in many shapes in accordance with the design and dimensions given on the building plan. A typical semi-circular arch is shown in Fig. 8-16A.

The arch center for this opening would be made up with ribs of either 1 inch or 2 inch lumber depending on the span of the arch. Ties are fastened across the bottom of the center and

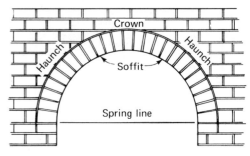

A. Semi-circular arch

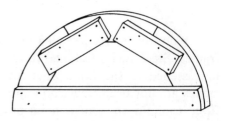

FIG. 8-16. Arch Center. B. Cleats on rib

supporting struts are put in to support the upper portion of the rib. The rib assembly is held together by cleats that have been nailed across the back of the mitered joints (see Fig. 8-16B).

The narrow strips that actually support the masonry are called lagging. They are usually made from 1 by 2 material cut to a length that will keep the outer ribs about $\frac{1}{2}$ inch back from the face of the brick wall. Making the arch center narrower than the wall facilitates finishing the mortar joints in the masonry.

The arch center is placed in the proper location and supported on shores. These shores are usually made up from 4 by 4 lumber cut to proper length and adjusted to final elevation with wood wedges placed below the shore. The shores are braced tightly against the existing masonry, and if necessary to maintain stability, diagonal braces should also be installed.

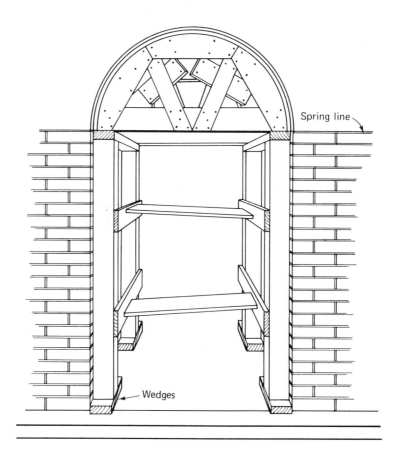

C. Arch center in place

FIG. 8-16 (continued).

Proper erection and bracing of the arch center will keep it from overturning while the masonry arch is being built (see Fig. 8-16C).

INSTALLING WINDOW BUCKS

A window buck is a rough frame installed in a masonry wall to maintain the proper size opening in the wall for a finished unit of the same overall size (see Fig. 8-17). The window buck is usually made of nominal 2 inch thick lumber of sufficient width to accommodate the wall

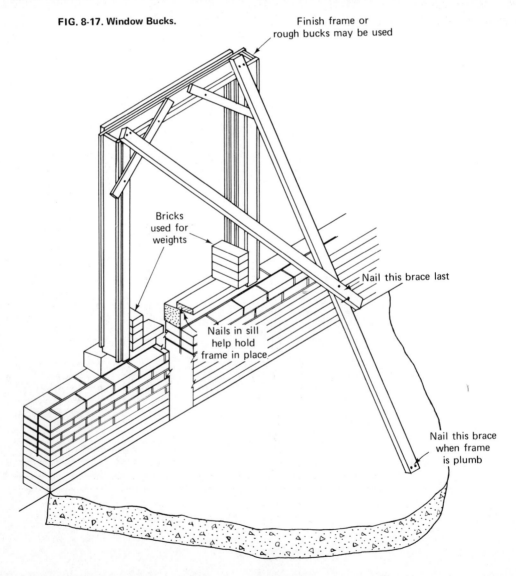

FIG. 8-17. Window Bucks.

thickness. Most window bucks are made from 2 by 8 or 2 by 10 planks.

The buck is made approximately $\frac{1}{4}$ inch larger in each direction than the overall size of the finished unit. This allows about $\frac{1}{8}$ inch clearance on each side around the finished frame and makes it easy to install. The buck is made with a sufficient number of spreaders and is braced diagonally to maintain proper shape.

Window bucks are installed when the masonry wall has reached the height of the bottom of the sill. Close cooperation between the masons and carpenters is required in order to maintain a steady work flow and also to locate the window bucks properly.

The bucks must be placed very carefully along the layout marks when the mortar is green to avoid breaking the joints. After the buck is set on the masonry wall, it must be braced laterally to keep it plumb. Work on the masonry wall can continue after the buck is in place with the mason using the buck as a guide for making the window opening.

On some jobs on which there is little or no danger to the window frame, the finished frames are installed instead of the window buck. The setting procedure is the same as for wood bucks, and the frame is held in place by masonry anchors fastened to it. These masonry anchors are embedded in the mortar joints as the wall is built.

INTERIOR PARTITIONS

Interior partitions may be made from wood or metal studs with gypsum wallboard or lath and plaster surfaces. The location of the partitions is marked out in accordance with the dimensions on the plans. Chalk lines are snapped on the floor between locating marks, and wall plates are fastened to the floor along these lines.

The location of the various door or window openings may be marked on the plates, or they may be marked directly on the floor. The framing members are placed in accordance with the layout marks.

Wood Stud Partitions

The length of the regular studs may be determined by making a story pole. The story pole for studding is usually a straight 2 by 4 long enough to accommodate the finish ceiling height plus the thickness of all the framing lumber. For a 10 foot ceiling, the carpenter might select a 2 by 4 that is 12 feet or 14 feet long and mark the ceiling height and the thickness of the various framing members on the story pole. After allowances for all the framing members have been made, the length of the studs is determined by measuring the distance between the top and bottom plates (see Fig. 8-18).

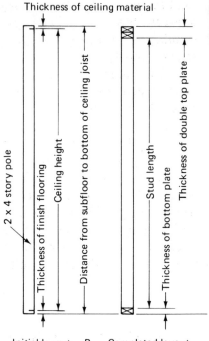

FIG. 8-18. Stud Length Story Pole.   A. Initial layout   B. Completed layout

When door and window openings are made in the wall, it becomes necessary to remove some of the regular studs. To support the framework above the opening, special framing members are installed. For doors, these framing members include headers, cripple studs, shoulder studs, and door studs (see Fig. 8-19). In some localities shoulder studs are known as cripples or trimmers. In addition to the above, window openings also require a rough sill and cripple studs between the rough sill and the sole plate.

Door Openings. To frame a rough opening for a door, the carpenter must know the rough opening size. This rough

INTERIOR PARTITIONS 191

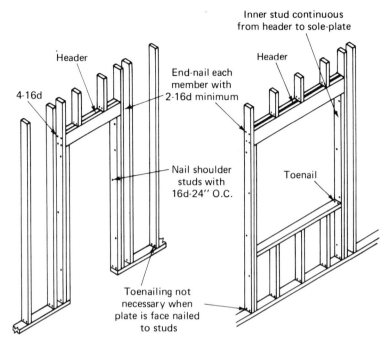

FIG. 8-19. Typical Door and Window Opening.

opening is the sum of the door size, the thickness of the frame on each side of the door, plus $\frac{1}{4}$ inch to $\frac{1}{2}$ inch framing allowance on each side of the frame. For interior door openings, a $\frac{1}{4}$ inch framing allowance is made on each side of the door jamb to take up irregularities between the rough framing lumber and the finished door frame.

The rough opening for a door 3 feet wide can be determined by adding all the necessary items as follows:

Door width	36"
Frame thickness	$1\frac{1}{2}$" ($\frac{3}{4}$" each side)
Framing allowance	1" ($\frac{1}{2}$" each side)
	$38\frac{1}{2}$" Rough Opening Width

The plate layout for this opening is shown in Fig. 8-20. Notice that the regular 16 inch O.C. marks are carried straight through and are not altered because of the door layout. The

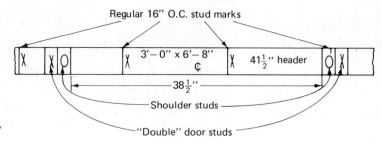

FIG. 8-20. Plate Layout for 36" Wide Door.

center of the opening is located on the plate in accordance with dimensions given on the plan, and one-half the rough opening distance is marked off in each direction. After the size of the opening is marked, the shoulder studs are drawn in. Notice that they are indicated by O's rather than X's. An additional regular stud is placed alongside the shoulder stud and is known as a "door stud." This door stud is the one to which the header is nailed.

In addition to marking the location of the framing members, the carpenter often marks in the size of the door and the length of the header. Because the header rests on the shoulder studs, 3 inches ($1\frac{1}{2}$ inches on each side) must be added to the rough opening size to get the header length. The header in this example must be $41\frac{1}{2}$ inches long. Marking the header length on the plate in this manner saves time in cutting and fabricating, because the workmen do not have to use their rules to measure each opening as they come to it.

Header Size

The size of header needed will vary with the width of the rough opening. Local building codes have requirements concerning header size, and a check of the various codes will show that there are variations in the requirements among the different codes. Therefore, the carpenter must know the building code requirements for the areas in which he works.

Table 8-1 gives the maximum spans for wall headers in outside walls and bearing partitions. These spans are acceptable as a general rule, and where there are no other code requirements they can be used with confidence. Partitions in stores are seldom bearing partitions. As a result, the header size over the opening in non-bearing walls may be reduced from that shown in Table 8-1.

Table 8-1. Maximum Spans for Wall Headers, Outside Walls, and Bearing Partitions

HEADER SIZE	MAXIMUM SPAN
2 — 2 by 4 on edge	4'0"
2 — 2 by 6 on edge	5'6"
2 — 2 by 8 on edge	7'6"
2 — 2 by 10 on edge	9'0"
2 — 2 by 12 on edge	11'0"

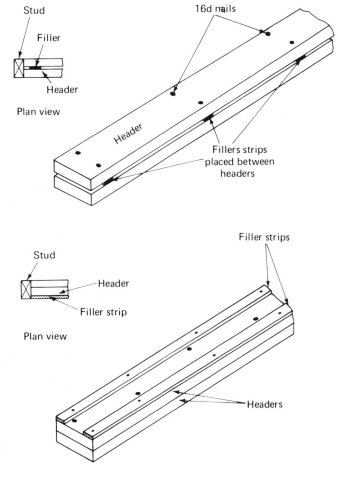

FIG. 8-21. Wall Headers with Filler Strips.

STOREFRONT CONSTRUCTION AND FINISHING

If a double 2 inch header is used, its combined thickness is only 3 inches ($1\frac{1}{2}$ inch + $1\frac{1}{2}$ inch), but the 2 by 4 wall is $3\frac{1}{2}$ inches wide. Therefore, the $\frac{1}{2}$ inch difference must be made up by a filler strip. This filler strip is usually made from $\frac{1}{2}$ inch plywood, or specially made rippings. The filler is usually placed between the header members, and 16d nails are driven through the headers and filler strips to form the header. The header is nailed together in this manner so that it will act as a unit (one piece) when it is installed in the wall.

An alternative method of forming the added thickness is to place the filler strips on the outside of the header after the header members are nailed together (see Fig. 8-21). This method is said to provide added header stiffness because of the friction between the header members.

Rough Opening Height

The distance from the subfloor to the bottom of the wall header is the same for most door and window openings and is governed by the height of the door frame. The height of the opening can be calculated mathematically, or it can be determined by additional layouts on the story pole.

To lay out the height on the story pole, the length of the door opening in the frame is first marked off on the story pole (see Fig. 8-22). For a standard 6 foot 8 inch door, this distance is $80\frac{1}{2}$ inches. This allows 80 inches for the door and $\frac{1}{2}$ inch for the threshold, which is placed below the door. If more distance between the floor and bottom of the door is required, the necessary amount is added to the length of the door to determine the height of the finished opening. Next, the thickness of the head jamb of the door frame is marked in. Head jambs are usually $\frac{3}{4}$ inch thick. Finally, a framing allowance of $\frac{1}{2}$ inch is marked in, and the location of the bottom of the wall header has been established.

In checking the layout, in this case the distance from the subfloor to the header is $82\frac{1}{2}$ inches. The $82\frac{1}{2}$ inches include an allowance for the thickness of the finish floor. In most cases this is also the distance from the floor to the bottom of window headers. The length of the shoulder stud that runs from the bottom plate to the header can be found by subtracting the plate thickness from $82\frac{1}{2}$ inches. Using a $1\frac{1}{2}$ inch thick plate, we would find the shoulder studs to be 81 inches long.

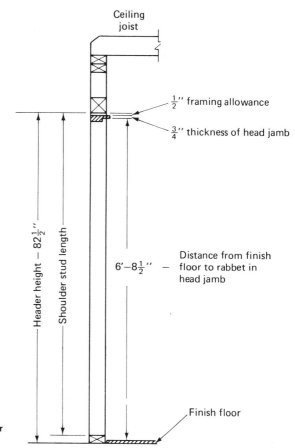

FIG. 8-22. Locating Header Height on Story Pole.

Now that the location of the bottom of the header has been established, the height of the header can be drawn in to determine the length of the cripple studs above the header (see Fig. 8-23). Notice that each different size of lumber header is drawn in and each line is labeled. The length of the cripples or blocks, as they are sometimes called, may also be marked on the story pole.

Window Openings. The layout of the rough opening for a window is similar to that for a door. The carpenter must know the size of the frame and add a $\frac{1}{4}$ inch to $\frac{1}{2}$ inch framing allowance on each side to determine the rough opening size.

Because of the wide variety of window frames available and the varying construction of these frames, it is recommended

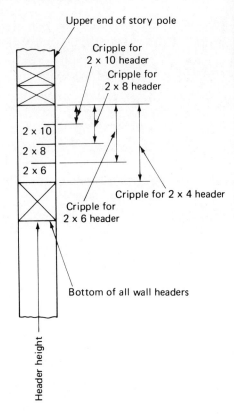

FIG. 8-23. Story Pole Layout for Header Cripple Studs.

that the carpenter obtain catalogs from the manufacturer whose frames he is using. These catalogs list the size of the frames and the rough opening needed for each different glass size. Reference to these catalogs can prevent mistakes in rough opening size and save the carpenter a lot of extra unproductive work.

The length of the cripple studs below the window opening may be determined by making additional layouts on the story pole. Starting at the bottom of the header, the carpenter measures down a distance equal to the rough opening height for the window frame and places a mark on the story pole (see Fig. 8-24). Next, the thickness of the rough sill is drawn in and the length of the cripple studs may be measured off between the bottom plate and the rough sill.

If there are many different window heights drawn in on the story pole, each is labeled and the length of the cripple studs are marked in for future reference.

INTERIOR PARTITIONS 197

Layout of Wall Plates. The location of all regular studs, corner posts, door and window openings, and intersecting partitions and backing studs is marked off on the top and bottom plates.

Marks are made on the floor, which establish the edge of the 2 by 4 plate, and chalk lines are snapped on the floor between the marks. This is the line along which the plate will be nailed.

If the wall runs at right angles to the joists, the studs should be placed directly above the joists or directly at the edge of the joists. Studs that are placed directly over the joists have the full support of the joists, and studs that are placed at the edge of the joists get sufficient support. An important factor to remember is that if studs are placed in this manner, there is always sufficient room for heating pipes placed between the joists to be turned up and run into the walls where required.

Using straight 2 by 4's for plates, the carpenter starts his layout so that the regular studs are positioned at the joists; but if the walls are erected on a concrete floor, he starts his layout by placing a corner post at the end of the wall and measuring in $15\frac{1}{4}$ inches to the side of the first stud (see Fig. 8-25). By starting the layout in this manner and continuing 16 inches O.C. from the first mark, the distance from the end of the wall to the center of the third stud from the corner post will be 48 inches. This type of layout makes it possible to use 48 inch by 96 inch or 48 inch by 144 inch sheets of wallboard and all the end joints will fall at the center of the stud.

Following the layout for all the regularly spaced studs, the intersecting walls are located in accordance with the plans and the backing studs are marked in. The location of the doors and windows is also marked in according to the dimensions on the plan, and the rough openings, shoulder studs, and double studs are indicated on the plates.

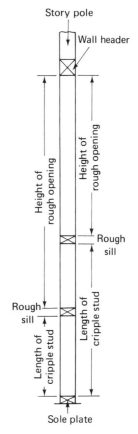

Note: Story pole may have layouts for many different sizes of windows — each layout should be indentified

FIG. 8-24. Story Pole Layout — Cripple Studs.

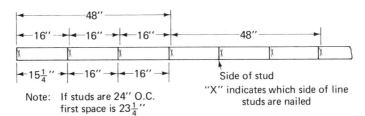

FIG. 8-25. Starting Plate Layout.

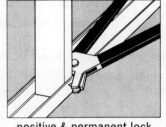

positive & permanent lock   pierces & folds light metal

FIG. 8-26. Metal Crimp Fastener. Courtesy United States Gypsum Co.

Metal Stud Partitions

If partitions are built of metal studs, a metal runner or track is fastened to the floor and ceiling along previously established layout lines, and studs cut to proper length are placed in the runners. The studs are held in place with special self-drilling and tapping screws, or they may be held in place with a metal lock fastener that punches the plate and stud and crimps them together (see Fig. 8-26).

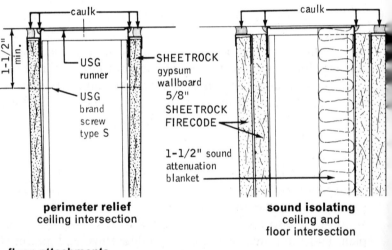
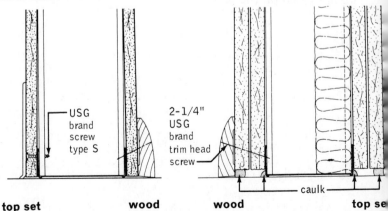

FIG. 8-27. Metal Stud Wall Details. Courtesy United States Gypsum Co.

INTERIOR PARTITIONS

Stud spacing is usually either 16 inches or 24 inches O.C. and is regulated by the thickness of the wallboard used on the partition. Some typical details showing construction at floor and ceiling junctions is shown in Fig. 8-27. Applying the wallboard will take place some time after the framing and all the mechanical work is completed. Gypsum wallboard is fastened to the studs with special bugle head screws. Wood trim is fastened with trim head screws (see Fig. 8-28).

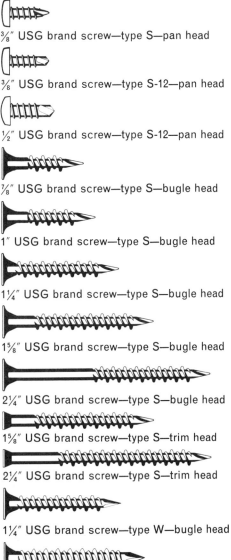

3/8" USG brand screw—type S—pan head

3/8" USG brand screw—type S-12—pan head

1/2" USG brand screw—type S-12—pan head

7/8" USG brand screw—type S—bugle head

1" USG brand screw—type S—bugle head

1 1/4" USG brand screw—type S—bugle head

1 5/8" USG brand screw—type S—bugle head

2 1/4" USG brand screw—type S—bugle head

1 5/8" USG brand screw—type S—trim head

2 1/4" USG brand screw—type S—trim head

1 1/4" USG brand screw—type W—bugle head

1 1/2" USG brand screw—type G—bugle head

FIG. 8-28. Screws for Drywall. Courtesy United States Gypsum Co.

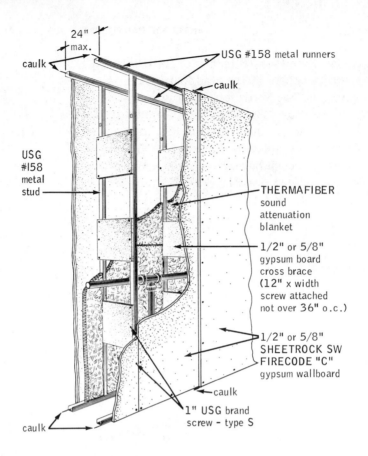

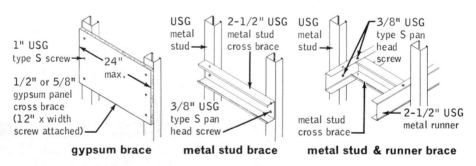

FIG. 8-29. Pipe Chase Wall. Courtesy United State Gypsum Co.

By using the proper combination of runners and studs, it possible to frame almost any type of wall (see Fig. 8-29).

Furring

Furring is used over wood walls to increase wall thickness, provide a flat nailing surface if the wall is irregular, and also for some types of soundproofing. On masonry walls, it may be used to provide a nailing surface, straighten irregular walls, or both.

Furring may be made from straight 1 by 2, 1 by 3, 1 by 4 boards or from 2 by 2 or 2 by 4 material. Metal furring is also available.

Furring used to increase wall thickness is simply nailed to the existing studs or masonry wall. If the wall surface is irregular, the furring must be shimmed to make it straight (see Fig. 8-30). Wood shingles are placed between the wall and furring and adjusted until the furring is brought into alignment with a straightedge or a taut line. Nails are then driven through the furring and shingles, and the furring is rechecked after the nails are in place. In some cases it will be necessary to readjust the shingles after nailing.

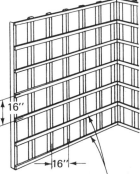

1" x 2" horizontal furring strips — shimmed as necessary to obtain flat and true plane

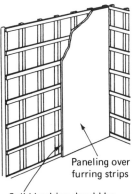

Paneling over furring strips

Solid backing should be provided behind edge joints

FIG. 8-30. Wall Furring.

Resilient furring channels are used to provide a degree of soundproofing and to separate the wallboard from movement in the structural framework. These channels are applied at right angles to the framing members and are placed 24 inches O.C. They may be fastened to wall framing with screws or nails but must be fastened to the ceiling framing with screws (see Fig. 8-31).

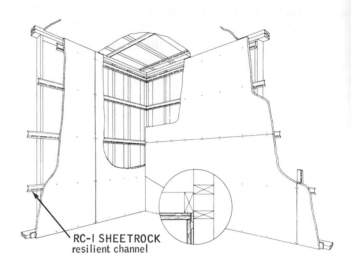

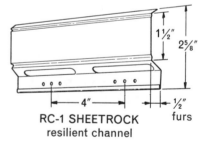

FIG. 8-31. Resilient Furring. Courtesy United States Gypsum Co.

Plaster Grounds

Plaster grounds serve as a guide or screed for the plasterer. They help the plasterer maintain proper wall thickness and straightness around the base of the wall and all door and window openings. Grounds may also be installed in other areas where a flat plastered surface is required.

Grounds are always installed by carpenters and are made from wood that will not be adversely affected by wet plaster. Most ground strips are $\frac{3}{4}$ inch by $\frac{3}{4}$ inch or $\frac{3}{4}$ inch by 1 inch, but other sizes may be used as required (see Fig. 8-32).

WALLBOARD AND PANELING

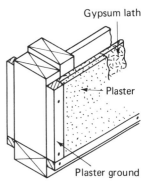

FIG. 8-32. Plaster Grounds.

The carpenter is called on to install various kinds of wallboards. Some of the most common materials are gypsum wallboard, hardboard paneling, plywood paneling, and solid wood paneling.

Installing Gypsum Wallboard

Gypsum wallboard may be applied vertically or horizontally. All vertical joints must occur on a framing member. If applied to wood studs, it may be fastened with screws, nails, or an adhesive nail-on combination (see Fig. 8-33). The use of an adhesive reduces the possibility of nail-popping caused by loosely nailed wallboard.

If the double-nail system is used to fasten gypsum board, a single nail is placed at the edges of the board at each stud or joist. Approximately every 12 inches between the edges two nails, 2 inches apart, are used to fasten the board (see Fig. 8-33). This system reduces nail popping by holding the wallboard tightly against the framing members.

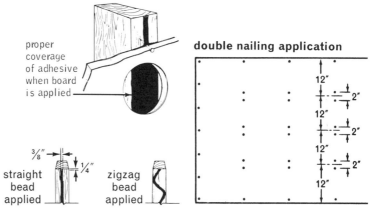

FIG. 8-33. Drywall Application. Courtesy United States Gypsum Co.

To reduce the possibility of cracks in the finished wall at interior corners, a floating interior corner is recommended (see Fig. 8-34). If the ceiling is made of gypsum wallboard, it is applied first. The wall material is butted tightly against the ceiling, and the first nails are placed about 8 inches from the ceiling. This method of nailing allows the building framework to move slightly without causing cracks in the finished gypsum corner.

On side walls, boards placed on the first wall are not nailed at the corner. They are held in place by the wallboard on the intersecting wall, which is nailed at the corner (see Fig. 8-34).

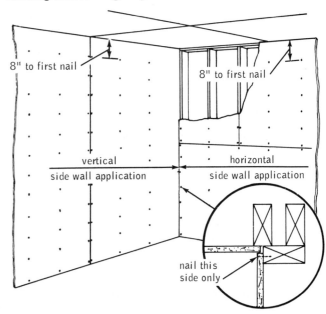

FIG. 8-34. Drywall Application. Courtesy United States Gypsum Co.

Installing Prefinished Paneling

Nearly all types of hardboard and plywood prefinished paneling may be fastened with adhesives, nails, staples, or screws. Adhesives do not leave holes in the paneling that later must be filled, but some method must be used to hold the panels in place until the adhesive sets.

If nails are used, they should penetrate at least $\frac{3}{4}$ inch into the studs or furring (see Fig. 8-35). Special hardened nails colored to match the paneling are available. If used, these nails

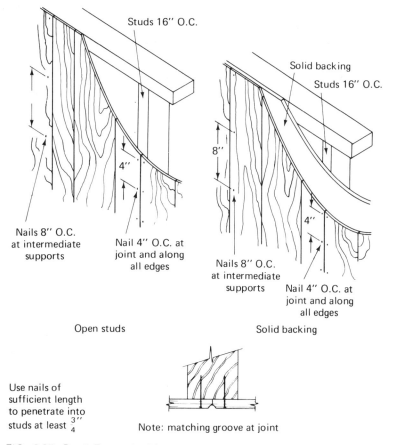

FIG. 8-35. Panel Fastened with Nails.

are driven flush to the surface of the paneling. The small flat heads of these nails blend into the panel and are inconspicuous. Therefore, setting the nail head below the surface of the panel is not required.

Caution must be used with colored panel nails because they are hardened and will break easily. Careless driving may cause them to break and fly and injure the eyes. Always wear safety glasses when driving hardened nails.

Finishing nails are usually placed in the grooves of the panels and set below the surface. At the edges of the panel, where it is not possible to place nails in the groove, the nails are placed approximately $\frac{1}{2}$ inch from the edge and are set below the panel surface. The nail heads are concealed by applying putty colored to match the paneling. Most panel manufacturers also manufacture a putty stick colored to match the various panels.

Various panel adhesives are available for installing prefinished panels. Most of these are packaged in cartridges for application with a caulking gun. Because of the many different types of adhesives available, it is recommended that specific manufacturers' instructions be followed.

All panel adhesives require clean and dry surfaces. In general, they are applied in a continuous bead along the center of the studding or furring, but for some types of adhesives an intermittent bead 3 inches long with a 6 inch space is acceptable on intermediate studs (see Fig. 8-36).

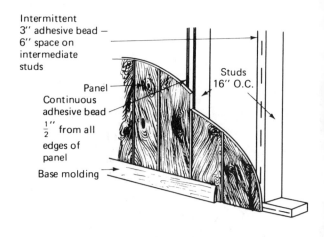

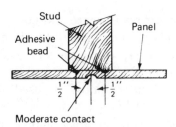

FIG. 8-36. Panel Fastened with Panel Adhesive.

After the adhesive is applied, the panels are set in place and pressed firmly to the framework to attain an initial bond with the adhesive. A minimum of two nails are placed at each end of the sheet to hold it in place while the adhesive cures.

After a short time, additional pressure is applied to all adhesive areas to obtain a final bond.

Contact cement is sometimes used to bond paneling to existing walls or directly to studs or furring. If contact cement is used, all surfaces must be clean and dry. A coat of cement is applied with a brush or roller to the surfaces that will be bonded together. After the cement on both surfaces is dry, the panel is positioned carefully and set in place. It bonds on contact with the prepared surface and cannot be moved. Therefore, extreme care must be taken in positioning the panel.

Mastic cements are used to fasten prefinished panels over existing walls of plaster, wallboard, or wood (see Fig. 8-37). The adhesive is applied over the entire area, which must be clean and dry. Panels are positioned and pressed against the mastic. The mastic will generally hold the panels in position without nails. However, if there are any unusual conditions, a small nail can be placed near the corner of the panel to hold it in position until the mastic sets.

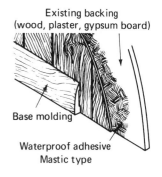

FIG. 8-37. Panel Fastened with Mastic Adhesive.

Panel Installation Procedure

The first panel to be installed is usually placed at an inside corner and held in a plumb position. If necessary, it is scribed to the adjacent wall (see Fig. 8-38). After scribing, the panel is placed on a pair of saw horses, cut with a fine-tooth saw, and trimmed with a block plane as necessary. After the necessary cutting is completed, the panel is put in place and checked for plumb with a carpenter's level. If ceiling and base trim will be

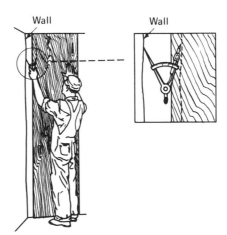

FIG. 8-38. Scribing a Panel.

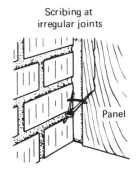

STOREFRONT CONSTRUCTION AND FINISHING 208

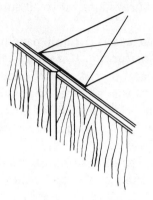

A. Standard V-groove joint

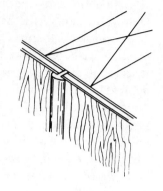

B. Joint molding

Butt joint
one panel scribed, other butted

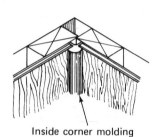

Inside corner molding

C. Inside corners

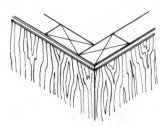

Mitered corner

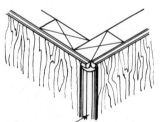

Outside corner molding

D. Outside corners

FIG. 8-39. Joint and Corner Finishing.

used, no further fitting is necessary, and the panel may be fastened in place.

Sometimes no moldings are used at the ceiling line. Then it is also necessary to scribe the panel to the ceiling after the panel is scribed and fitted to the first corner. It is placed in the corner and raised to the ceiling while the edge is maintained in a plumb position. The top edge is scribed to the ceiling and is cut and fitted in the same manner as the first edge. After the trimming and fitting is completed, the panel is placed in position, checked for alignment, and fastened to the wall.

Most prefinished panels are joined by simple butt joints that are made inconspicuous by a V-groove (see Fig. 8-39A), but sometimes a joint molding (see Fig. 8-39B) is used. This molding is placed over the edge of the panel after the panel is installed. The molding is then fastened to the wall framework by nailing through its exposed leg.

Inside corner joints are usually scribed and butted. However, inside corner moldings are available for use with many different types of panels (see Fig. 8-39C). Many of these moldings are finished to match the type of paneling being installed.

Outside corners of prefinished paneling may be mitered as shown in Fig. 8-39D. The mitered corner gives a pleasing appearance. Unfortunately, it can withstand only limited abuse and should be used only if it is not subject to heavy traffic. In making the mitered corner, a straightedge should be clamped in place to the back of the panel to provide a guide for the portable power saw. This procedure assures a straight cut and a well-fitting mitered corner.

Metal, plastic, and wood outside corners are available for various types of prefinished panels. The metal type is installed before the second corner panel is put in place (see Fig. 8-39D). Wood corner beads prefinished to match the paneling are applied after the panels are in place. Each type of molding offers a pleasing appearance and provides a durable corner that can withstand more abuse than the mitered corner.

ACOUSTICAL CEILINGS

Acoustical ceilings offer a wide variety of pleasing appearances and serve a number of useful functions. By using different materials, the ceiling can be made to absorb greater or lesser amounts of sound. The surface texture can be varied to alter the appearance, and the texture can be used to hide ventilation openings. Some acoustical ceilings serve as a fire barrier and may have a 1, 2, or 3 hour fire rating.

Types of Acoustical Ceilings

Acoustical ceiling tiles and panels are manufactured from a variety of materials that give them different characteristics. One type of ceiling material is made up entirely of noncombustible mineral fibers and other products. Other tiles are made of glass fiber, metal, or wood products. Standard tiles are 12 inches square and usually $\frac{5}{8}$ inch thick. Ceiling boards are manufactured in thicknesses from $\frac{1}{2}$ inch to 3 inches depending on size and material used. Most standard ceiling boards are either 2 feet by 2 feet or 2 feet by 4 feet in size, but some ceiling boards are made 2 feet 6 inches by 5 feet.

Some acoustical ceiling materials are affected by high humidity. The manufacturer's recommendations should be followed in regard to building conditions during the time of ceiling installation. In general, the window areas should be glazed, and all exterior doors should be installed before the ceiling material is delivered. The building area should be kept at a temperature and humidity as near equal to that which will prevail when the building is occupied.

Methods of Installation

Acoustical materials can be installed by various methods, including adhesive application; nail, screw, or staple application; and mechanical suspension using a concealed or exposed grid system of various types. The type of grid system used will depend on the type of tile or panel and the type of ceiling construction. Before any ceiling work can be done, all the necessary plumbing, electrical work, and heating and air conditioning work that will be enclosed in the space between the acoustical ceiling and the bottom of the floor slab must be completed. After checking the plans and specifications against job conditions, the carpenter will proceed to lay out the ceiling and locate the starting lines. If there is a ceiling layout included with the plans, that layout is followed.

Horizontal Layout. With few exceptions attributable to job conditions, the tiles should be laid out in a manner so that the border tiles will not measure less than $\frac{1}{2}$ the width of the tile. Some mechanics will make their layout on the floor and transfer the starting lines to the ceiling after they have been located. Other mechanics may do the layout by locating center lines on the ceiling or by mathematical means.

Many times, determining the width of the edge row of tiles and the location of the starting line is easily accomplished mathematically using the following procedure:

1. Find center of room in each direction by dividing overall dimensions by 2.
   NOTE: Room must be rectangular or square. For rooms with wings the tile is usually laid out for the larger area and the starting lines are extended into the wings.
2. Divide the half width of the room by the width of the tile. If the result is a whole number or a whole number and a fraction of more than $\frac{1}{2}$, the starting line is at the center of the room. If the fraction is less than $\frac{1}{2}$, the starting line is moved so that the room center line runs through the center of the tile (see Fig. 8-40). The width of the edge tile is found by dividing the distance from the wall to the starting line by the width of the tile. The whole number is the number of full tiles in that distance, and the remainder is the width of the edge tiles.
3. Divide the half length of the room by the length of the tile. If the result is a whole number or a whole number and a fraction of more than $\frac{1}{2}$, the starting line is at the center of the room. If the fraction is less than $\frac{1}{2}$, the starting line is moved so that the room center line runs through the center of the tile (see Fig. 8-40). The length of the edge tiles in this direction is found by dividing the distance from the wall to the starting line by the length of the tile. The whole number is the number of full tiles in that distance, and the remainder is the length of the edge tiles.

STOREFRONT CONSTRUCTION AND FINISHING 212

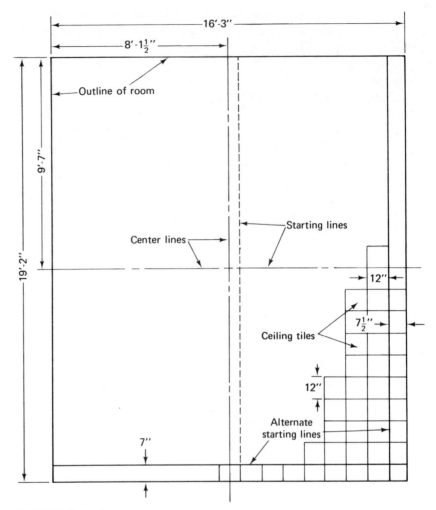

FIG. 8-40. Ceiling Layout.

EXAMPLE

Find the center and starting lines for 12″ by 12″ tiles in a room 16′3″ by 19′2″ (Fig. 8-40).

1a. 16′3″ = 195″ ÷ 2 = 97.5″ from wall to center line

2a. 97.5 ÷ 12 = $8\frac{1}{8}$ blocks

Fraction is less than $\frac{1}{2}$. Therefore, starting line is moved 6″ from the center line.

Find the width of the edge tiles.

91.5″ ÷ 12″ = 7 full tiles and $7\frac{1}{2}$″ edge tile

1b. 19'2" = 230" ÷ 2 = 115" from wall to center line

2b. 115 ÷ 12 = $9\frac{7}{12}$ blocks

Fraction is more than $\frac{1}{2}$. Center line in this direction represents starting line.
Find the length of the edge tiles.
115" ÷ 12" = 9 full tiles and 7" edge tile

After the size of the edge tiles is determined, the layout is marked on the walls or ceiling, and the mechanic is ready to establish his starting line for the tiles or for the suspended grid system, if used.

Vertical Layout. Good practice requires that acoustical ceilings be within $\frac{1}{8}$ inch of level in 12 feet. Therefore, the layout for the carrying channels and perimeter angles must be carefully done. Working from some pre-established point, the mechanic will use a water level or a transit to establish the ceiling line at the corners of the room and at intermediate points on long walls.

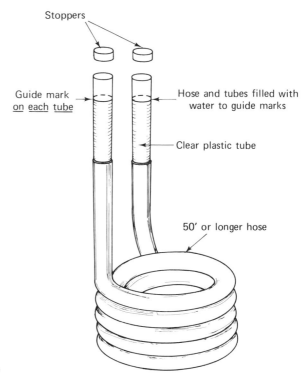

FIG. 8-41. Water Level.

The water level is generally more suitable for ceiling work as it can be used in obstructed areas and does not depend on an open line of sight. A water level consists of a clear plastic hose or a rubber hose fitted with a clear plastic tube at each end (see Fig. 8-41). The hose is filled with water, and a stopper is fitted to each end to prevent loss of water. If the height of the water is marked on the tube at each end of the hose, it is possible for one person to locate all ceiling elevation marks around the perimeter of the room (see Fig. 8-42).

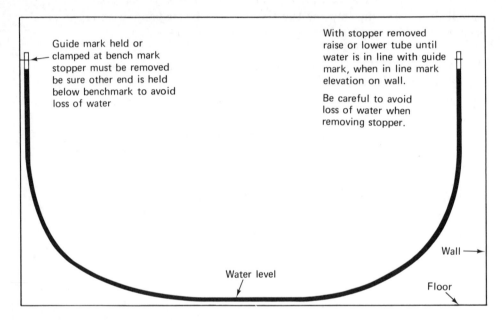

FIG. 8-42. Using Water Level.

Some of the larger ceiling contractors are using lasers to establish the level of a ceiling. The laser is a rather expensive precision tool, but it offers the advantage of establishing level lines over a large area. If a laser is to be used to establish level lines the manufacturer's manual should be studied and instructions in the use of the laser be carefully followed.

Tiles Installed With Adhesive. Tiles installed with adhesives can be applied over most clean and dry surfaces. The adhesive manufacturers' recommendations should be carefully followed, but in general one dab of adhesive about the size of a walnut

ACOUSTICAL CEILINGS   215

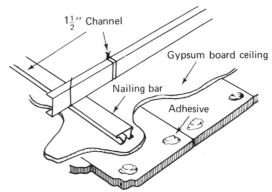

FIG. 8-43. Adhesive Application. Courtesy Owens-Corning Fiberglas Corp.

placed about 3 inches from each corner of a 12 inch by 12 inch tile is sufficient (see Fig. 8-43).

When working with tiles, hands must be clean. Rubbing clean hands with talcum powder helps keep them clean, but in many cases it is desirable to wear white cotton gloves when applying tiles. Tiles are set in place with a sliding motion, and care must be taken to press all tiles evenly in place. If the ceiling surface is uneven, the irregularities can be taken up by applying more or less adhesive to compensate for the unevenness.

It is good practice to work from more than one box of tiles to avoid large areas of slightly different colors. The tile manufacturer's installation instructions are inside the box. They should be studied before starting the installation and followed carefully.

Concealed Z-Spline Installation. The concealed Z-spline provides a sturdy framework on which to install grooved ceiling tile. The Z-splines are attached to suspended $1\frac{1}{2}$ inch channels by special clips (see Fig. 8-44). The channels are suspended

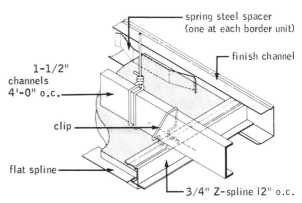

FIG. 8-44. Concealed Z-spline. Courtesy United States Gypsum Co.

from the underside of the floor slab by #9 S.W.G., wire which is attached to inserts set in the concrete.

To install a ceiling of this type, the elevation of the ceiling is established around the perimeter of the room with a water level or other device, and finish channels are installed around the perimeter of the room. The elevation of the bottom of the $1\frac{1}{2}$ inch carrying channels is then established and channels are located on 4 foot centers. Carrying wires are tied to inserts in the concrete ceiling every 4 feet along the length of the carrying channel. These wires are bent at the elevation of the bottom of the carrying channel and tied around the channel (see Fig. 8-44). The channel must be carefully tied to maintain a straight level ceiling. Wires that are bent and tied too low or too high will leave irregularities in the ceiling that will be noticeable under varying lighting conditions.

The Z-splines are attached to the channels with special clips. The first Z-spline is fastened along the starting line at the center of the room. After the Z-spline is installed, the groove of the tiles is carefully placed on the spline and flat splines are placed between the tiles in the grooves at right angles to the Z-spline (see Fig. 8-44). As ten or twelve tiles are set in place, additional Z-splines are placed in the grooves of the hanging tiles and clipped into place.

Exposed Z-spline suspension systems are installed in the same manner as concealed systems. The only difference is the size of the Z-spline, and the manner in which the tiles are supported. The tile boards rest on the horizontal leg of the Z-spline, and flat splines are placed in the groove between the tiles (see Fig. 8-45).

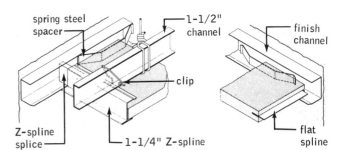

FIG. 8-45. Z-spline Suspension. Courtesy United States Gypsum Co.

Exposed Grid System. The exposed grid is a direct suspension system composed of main T's and cross T's. The main T's are suspended by wire fastened to anchors in the concrete ceiling at

4 foot intervals. Main T's are usually placed on 4 foot centers, and cross T's are placed on 2 foot centers between main T's (see Fig. 8-46). A $\frac{3}{4}$ inch by $\frac{3}{4}$ inch edge angle is used to support the tiles at the perimeter of the room.

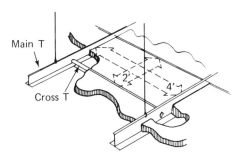

FIG. 8-46. Exposed Grid. Courtesy Owens-Corning Fiberglas Corp.

Layout for the exposed grid system is made in the same manner as for the Z-spline system. The main difference being that the main T's take the place of the channel used on the Z-spline system, and flat splines are replaced by cross T's. As in any ceiling system, the manufacturers' recommendations should be followed to attain a good completed job.

REVIEW QUESTIONS

1. Why is knowledge of materials and construction details important to the builder of storefronts?
2. Outline the procedure for making the storefront wall layout.
3. What is the purpose of a storypole?
4. Outline a procedure for fabricating storefront walls.
5. How are window display areas built?
6. Outline a procedure for installing wood frames.
7. List some of the materials used for storefronts.
8. What precautions should be taken when installing bevel siding?
9. What precautions should be taken when installing tongue and groove siding?
10. What precautions should be taken when installing board and batten siding?

11. What precautions should be taken when installing plywood sidings?
12. Outline a procedure for installing porcelain enamel panels.
13. How are storefronts prepared for large metal framed glass openings?
14. Outline a procedure for installing steel door bucks.
15. What are arch centers?
16. Outline a procedure for installing window bucks.
17. Outline a procedure for building wood stud partitions.
18. Outline a procedure for building metal stud partitions.
19. Outline a procedure for installing furring strips.
20. Outline a procedure for installing wallboard.
21. Outline a procedure for installing prefinished paneling.
22. Outline a procedure for horizontal layout of acoustical ceilings.
23. Outline a procedure for vertical layout of acoustical ceilings.

MOVABLE PARTITIONS
CHAPTER NINE

Movable partitions are used in offices, retail stores, manufacturing plants, hospitals, schools, and any other buildings in which there is an occasional need to rearrange the location of partitions. These reusable partitions are made in various size panels and may be made from a number of different materials. Some are made entirely of wood and have special fastening and connecting hardware. Another type is made entirely of steel and has a porcelain enamel or other painted finish. The steel partition systems also have their own special fastening and connecting hardware.

Movable gypsum partitions are produced by a number of manufacturers in various designs and heights. Like other movable partitions, these partitions are a complete system with special runners, studs, wall panels, trim, and hardware. Some of these partitions are made of panels with a solid gypsum core, or a wood fiber core, whereas others utilize a metal stud system and gypsum wallboard. Some of the wall panels are available with a plastic laminate or vinyl surface.

The terms "demountable partition" and "movable partition" can be applied to the same wall system. However, some manufacturers differentiate between movable and demountable partitions by the manner in which they are built and dismantled. By definition, the movable partition is built or moved in sections 2 feet to 4 feet wide, but the demountable partitions are dismantled into component parts that can be reassembled. As used in this chapter, the term *movable partition* will apply to all types of movable and demountable partitions.

Movable partitions are made in various heights. They can be installed from floor to ceiling, but they are also available in cornice height, approximately 7 feet high; and counter height, approximately 4 feet high.

PARTITION LAYOUT

Before beginning any partition layout, make a careful and thorough study of the floor plan and framing details. The method of fastening partition sections to the floor, ceiling, and intersecting walls should be carefully examined so that the partition layout can be made without overlooking important framing details.

After studying the plans, the dimensions of the work area should be checked against the plan dimensions. Any deviations between work area and plan dimensions should be written on the plan so that any necessary adjustments may be considered.

After the study of the plans, the checking of dimensions, and the necessary adjustments have been completed, the partitions may be located on the floor. The usual procedure is to start measuring from a wall or column and to mark the location of the partition runner in accordance with plan dimensions. After the partition location has been established at each end, chalk lines are snapped between the marks. Layout for all the walls is done in a similar manner.

After the various wall locations have been established, door and window openings are marked on the floor in preparation for locating floor runners, post anchors, and other framing members.

INSTALLING MOVABLE PANEL WALLS

The procedure followed in the installation of movable walls varies with the type of wall panel and fastening system. Working with a typical system using a floor and ceiling runner, the first step is to install the runners in accordance with lines established on the floor. After the runners are in place, installation of wall panels may begin. This is best accomplished by starting at a corner or door opening and working down the length of the wall.

As each panel is set in place, it is adjusted to the proper position and fastened with the appropriate devices. As the wall erection is completed, the necessary wiring and piping is installed in the open spaces provided in the wall system. Batten strips and floor and ceiling trim are installed after all work contained within the wall is completed.

Because panels made by different manufacturers have different connecting, fastening, and trim features, it is advisable to study the detail drawings supplied with the wall components. Some movable wall manufacturers also provide detailed assembly instructions with the walls. If these are provided, they should be studied before starting work on the wall in order to save time and costly mistakes.

DEMOUNTABLE PARTITIONS

Demountable gypsum wallboard partitions are built from steel studs, steel track, gypsum wallboard, and aluminum or steel trim. The gypsum panels may be $\frac{1}{2}$ inch or $\frac{5}{8}$ inch thick and covered with a vinyl material, or they may be plain panels that are painted before the trim is applied. Materials for this type of partition are available from a number of manufacturers.

Ceiling Height Partitions

Ceiling height demountable partitions may be made to include door openings and window panels. The location of the wall is marked on the floor and chalk lines are snapped in accordance with the plan layout. Door and window openings are also marked out before the installation of the various components begins. A typical wall elevation in Fig. 9-1 shows the various features that can be designed into this type of wall. The circled numbers refer to the framing details given in Fig. 9-2.

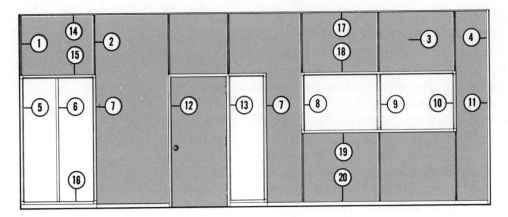

FIG. 9-1. Demountable Ceiling Height Wall. Courtesy Gold Bond Building Products, Division of National Gypsum Co.

Number 1 in Fig. 9-2 shows a typical stud placed against an intersecting wall. If this stud is not intersected by an opening, it is usually fastened to the intersecting wall and also to the floor and ceiling runners. The batten strips are installed after the wallboard is in place to hide the joint between the intersecting partitions.

Numbers 2 and 3 show the typical stud arrangement. Notice how the panel joints are staggered on opposite sides of the wall. Staggering the joints in this manner results in a stiffer wall. Studs are normally spaced 24 inches O.C. and are fastened to both the top and bottom runners with self-drilling and tapping screws.

If the end of a wall will be exposed, a trim cap can be placed over the end stud and wallboard as shown in Numbers 4 and 11. Care must be taken in placing the end stud perfectly plumb because the end trim will be plumb only if the stud is plumb.

Numbers 5, 6, and 7 in Fig. 9-2 show typical vertical framing around door height windows. The 1 by 4 wood strip in 5 takes the place of a stud and is fastened to the existing wall to provide support for the aluminum glazing channel. A similar glazing channel is used at 7, but it is applied over the metal stud after the wallboard is in place. The glazing divider bar is installed after the horizontal members are in place.

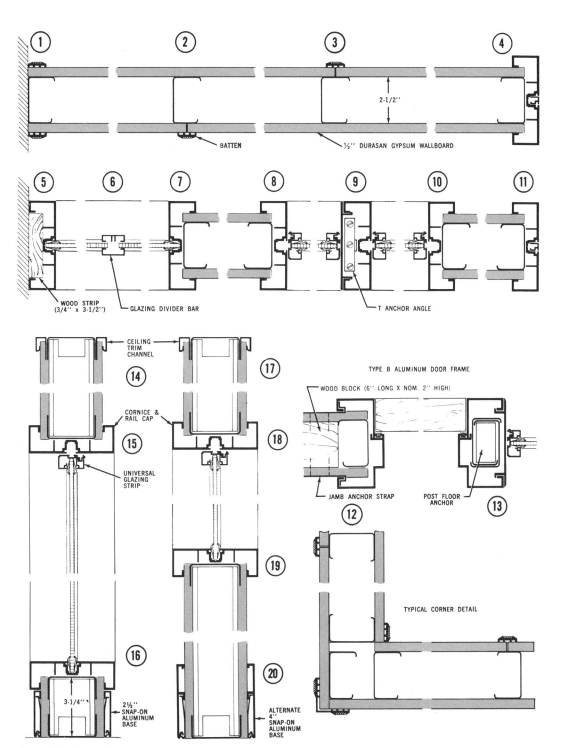

FIG. 9-2. Ceiling Height Wall — Details. Courtesy Gold Bond Building Products, Division of National Gypsum Co.

223

Details 8, 9, and 10 show typical vertical framing for smaller window openings. These vertical members are installed after the horizontal trim is in place. Each vertical is made from a typical rail cap or jamb unit and is fitted with the necessary glazing stops and filler cap.

Details 12 and 13 show the vertical framing for the door opening. Notice how wood blocking is placed within the wall at floor line to provide support for attaching door jamb anchors. Notice, too, that the jamb between the door and side light is fastened at the floor with a special post floor anchor.

Details 14 and 17 show the channel fastened to the ceiling. Studs are set inside the channel and fastened to it with self-drilling and tapping screws. The ceiling trim channel is set tight against the ceiling to hide the joint between the ceiling and wallboard. The wallboard is set into the trim channel.

The framing in Details 15, 18, and 19 is identical. A metal channel is used to hold stud spacing above or below the openings, and the cornice and rail cap is placed after the wallboard is installed. The required glazing stops are installed after the rail cap is in place.

The framing at the base in Details 16 and 20 shows a channel, which is fastened to the floor, and studs set in the channel. After the wallboard is in place, hardware for a snap-on base is installed and the baseboard is snapped in place to cover the joint between the wall and the floor.

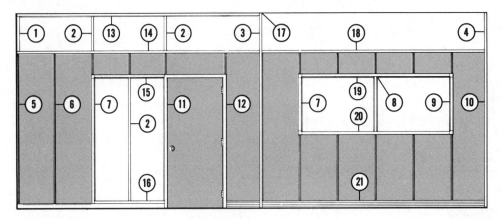

FIG. 9-3. Ceiling Height Wall — Elevation. Courtesy Gold Bond Building Products, Div. of National Gypsum Co.

Cornice Height Partitions

Cornice height partitions may be made up to include door openings and window panels. The area between the top of the wall and the ceiling may be left open for borrowed light and ventilation, or it may be glazed for borrowed light and privacy. A typical cornice height partition elevation in Fig. 9-3 shows the various features that can be designed into a wall of this type. The circled numbers refer to the framing details given in Fig. 9-4.

Numbers 1, 2, and 3 in Fig. 9-4 show typical vertical framing around borrowed light windows. The 1 by 4 strip in 1 takes the place of a stud and is fastened to the existing wall to provide support for the aluminum glazing channel. Detail 4 shows the cover plate on the glazing channel in the area above the wall. Details 5 and 6 show typical stud and batten application. Notice how the joints in the gypsum board are staggered to provide greater wall stiffness.

Details 7, 8, and 9 show typical vertical framing around window openings. These vertical members are installed after the horizontal trim is in place. Each vertical is made from a typical rail cap or jamb unit and is fitted with the necessary filler caps and glazing stops.

Detail 10 shows the vertical section of a typical floor-to-ceiling post. These posts, made of rail cap material, are fastened to both the floor and ceiling and are set a maximum of 12 feet on center. The stud in this assembly runs from the floor to the cornice rail.

Details 11 and 12 show the vertical framing around the door. Notice the post floor anchor on the window side of the door and the wood block on the floor at the other side of the door. These items are used to anchor the door frame securely.

Details 13, 14, 15, and 16 show typical framing around the window opening. Notice the top and bottom channel, which is used as a plate on each end of the short stud wall portion of the wall. The cornice and rail cap is installed over the channel after the gypsum panels are in place. The cornice and rail cap at the ceiling is held in place by a 1 by 4 fastened to the ceiling.

Detail 17 shows the use of a T-anchor angle to fasten floor to ceiling posts to the ceiling construction. An anchor of this type is required whenever floor to ceiling posts do not frame into a stud channel.

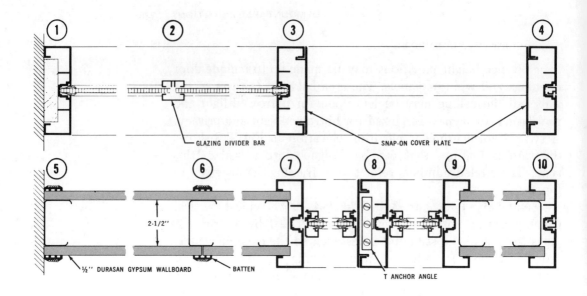
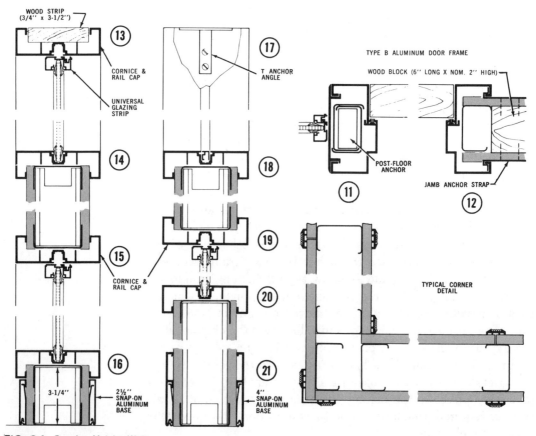

FIG. 9-4. Cornice Height Wall — Details. Courtesy Gold Bond Building Products, Div. of National Gypsum Co.

Rail Height Partitions

Rail height partitions are fabricated with open or closed base construction. If closed base construction is used, it is the same as for open base except for the location of the base board. The elevation of a typical rail height partition with an open base is illustrated in Fig. 9-5. The circled numbers refer to the framing details given in Fig. 9-6.

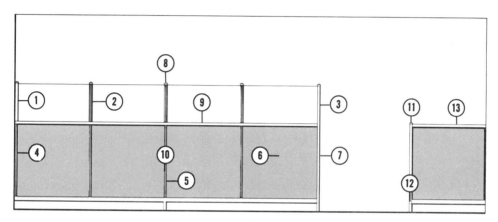

FIG. 9-5. Rail Height Wall — ELEVATION. Courtesy Gold Bond Building Products, Div. of National Gypsum Co.

The metal stud in Detail 4 is fastened to the existing wall, and the nested metal studs at Detail 5 are fastened to the post floor anchor. The stud at Detail 6 is fastened to the metal channels shown in Details 9 and 10. The nested studs in Detail 7 are fastened to a post floor anchor as at 5. The end of the partition is covered with a cornice and rail cap after the wallboard is in place.

The glazing post shown in Details 2, 8, and 9 is fastened to the cornice and rail cap before it is installed. All cornice and rail caps, glazing assemblies, post anchor caps, base, and battens are installed after the wallboard is in place on the metal framework.

Installing Demountable Gypsum Partitions

Following the floor plan, the location of the partition is marked on the floor and ceiling. Door and window openings are also marked along the partition layout. The channels, or tracks,

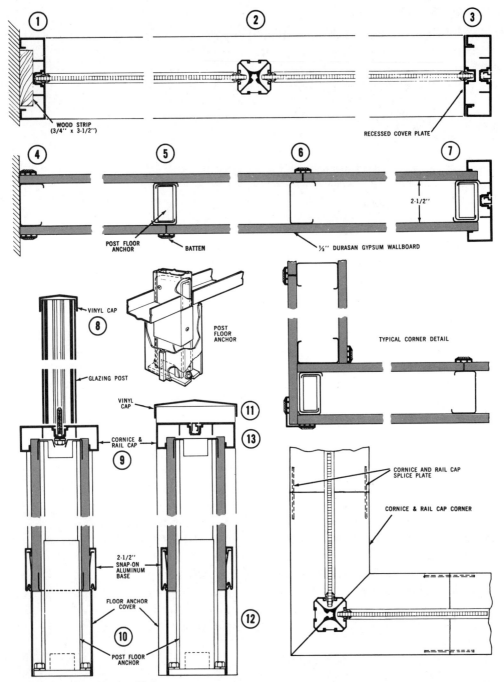

FIG. 9-6. Rail Height Wall — Details. Courtesy Gold Bond Building Products, Div. of National Gypsum Co.

as they are called, are fastened to the floor and ceiling in the manner required by the plans and specifications. Metal studs are then placed in the track and fastened with sheet metal screws or by crimping.

The necessary framing around door and window openings is made by cutting and nesting the studs and track as required. For maximum rigidity, two studs may be nested on each side of door and window openings in the manner shown in Detail 7, Fig. 9-6.

When the framing is complete, ceiling trim channels are applied on each side of the ceiling track. The gypsum wallboard is cut to size and fastened in place. The wallboard may be fastened with an adhesive and screws, but in some cases screws alone are used.

After the wallboard is applied, base clips and base trim are installed. Battens and window and door trim are also installed at this time. For cornice height partitions, it is necessary to install the cornice cap trim on top of the partition, and a post cover plate on the open side of the partition posts. A careful study of detailed plans provided with the partition system should help solve any problems concerning assembly and trim attachment.

REVIEW QUESTIONS

1. Where are movable partitions used?
2. Is there any difference between demountable and movable partitions?
3. What are cornice height partitions? Counter height partitions?
4. Outline a procedure for laying out movable partitions.
5. Outline a procedure for building ceiling height movable partitions.
6. Outline a procedure for building cornice height movable partitions.
7. Outline a procedure for building counter height movable partitions.
8. What precautions should be taken when handling prefinished wall panels?

CABINET AND FIXTURE WORK
CHAPTER TEN

A wide range of cabinets and fixtures is installed on commercial jobs. These cabinets may be made from wood, wood products, plastic, metal, or a combination of any of these materials. The types of cabinets and fixtures are almost endless and include wardrobes, clothes racks, display cases, storage cabinets, service counters, workbenches, and decorative fixtures.

The installation of fixtures is always custom work. Therefore, the carpenter must study the floor plans, elevations, and special details to determine the location of the various units and the manner in which they are to be installed.

DECORATIVE CORNICES

Decorative cornices are installed around the perimeter of retail stores, display rooms, meeting rooms, and other facilities. They are custom designed for the installation and therefore the number of designs is endless (see Fig. 10-1). Seldom will one design be repeated from building to building, but the methods of fitting and fastening cornices all fall into a general pattern.

DECORATIVE CORNICES   231

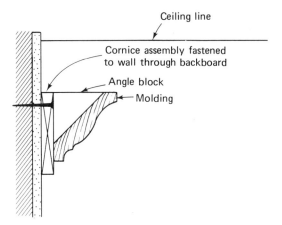

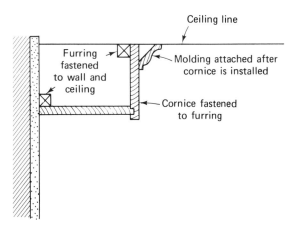

FIG. 10-1. Typical Cornices.

After checking the plans and determining the elevation of the cornice, the location is marked on the wall and with the use of a water level, or other leveling device, the height of the cornice is marked at various points around the specified perimeter. Chalk lines are snapped between the established marks, and the wall is ready for cornice attachment.

The cornice members are often partially prefitted and almost always prefinished. Therefore, extreme care in handling is important so that none of the prefinished parts becomes damaged. Prefitted parts may be held up to the chalk line and fastened in place. The type of fastener to be used will vary from job to job and is specified on the plans and in the job

CABINETS AND FIXTURE WORK   232

specifications. If the type of fastener is not specified, the carpenter must use good judgment and consider the type of wall construction to which the cornice will be fastened.

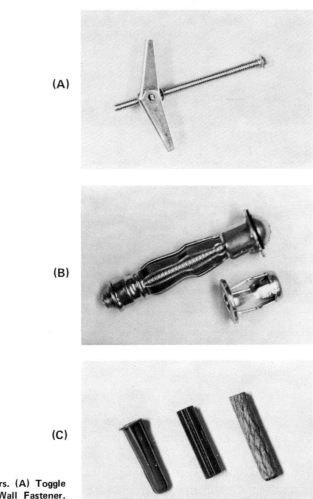

FIG. 10-2. Fasteners. (A) Toggle Bolt. (B) Hollow Wall Fastener. (C) Fiber and Plastic Screw Anchors.

On hollow walls, the cornice may be fastened with toggle bolts, hollow wall anchors, or expansive plastic plug anchors. If the cornice must be fastened to a masonry wall, plastic plugs, fiber screw anchors, or one of a variety of metal expansive anchors can be used. Self-drilling and tapping screws can be used to fasten the cornice to metal studs, but if they are not available holes can be drilled through the cornice into the metal studs and self-tapping screws used to fasten the cornice (see Fig. 10-2). Whenever possible, the fasteners should be placed so that they will not be visible. If the fasteners are visible, they must be set below the surface of the cornice and the holes filled to maintain an attractive appearance.

Fitting Inside Corners

Inside corners of molded cornices are mitered and coped. A simple miter cannot be used in most cases because the joint

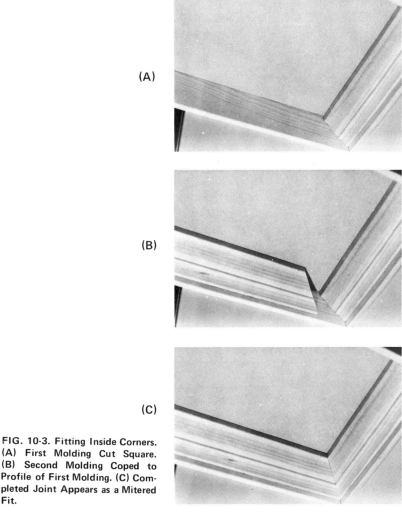

FIG. 10-3. Fitting Inside Corners. (A) First Molding Cut Square. (B) Second Molding Coped to Profile of First Molding. (C) Completed Joint Appears as a Mitered Fit.

has a tendency to open up. The more elaborate the cornice, the more difficult will be the job of fitting the inside corner. However, if the cornice is first mitered, the profile of the cornice can be followed with a coping saw or portable saber saw and the back wood removed (see Fig. 10-3).

The first section of cornice installed is cut square at the end and run into the corner so that the end fits tightly against the intersecting wall. The mitered and coped section is fitted against the first section and fastened in place (see Fig. 10-3). The final appearance is that of a mitered joint.

Whenever possible, avoid the need to cope both ends of a single cornice section because it is very difficult to get a good fit at each end. If the cornice arrangement is such that a coped joint is required at each end, it is usually better to install the cornice in two sections. Each corner can then be coped and fitted separately, and the end joint where the two sections come together can be easily fitted with a square cut or mitered ends.

Fitting Outside Corners

Outside corners are mitered. After inside corners are fitted, the cornice section is marked to length, and the outside end is

FIG. 10-4. Fitting Outside Corners.

mitered. The cornice must be mitered carefully because any deviation from a perfect miter will result in an open joint. Clamping the cornice in the miter box will help obtain a perfect cut. As the cornice sections are cut, they are fastened in place (see Fig. 10-4).

DISPLAY RACKS

The location of display racks — shelves, poles, and standards — is given on the floor plan and elevations. Various detail drawings show the manner in which the various parts are to be connected and fastened in place.

After studying the plans, the first step in installing display standards is to mark their location on the wall and floor as required. The standards are fastened in place with nails, screws, and anchors as specified. Some of the typical display rack hardware is shown in Fig. 10-5. By installing the proper brackets on the standards, the display rack can be made to accommodate flat or sloping shelves, clothes poles, dust covers, and decorative cornices as required.

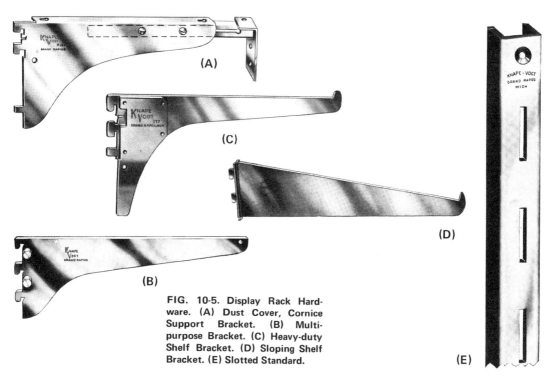

FIG. 10-5. Display Rack Hardware. (A) Dust Cover, Cornice Support Bracket. (B) Multipurpose Bracket. (C) Heavy-duty Shelf Bracket. (D) Sloping Shelf Bracket. (E) Slotted Standard.

CORNICE HEIGHT DISPLAY CASES

Cornice height display cases are generally shipped to the job as assembled units. When they arrive at the job, they are placed according to the floor plan and elevations. Before the unit is fastened in place, it is leveled by placing shim shingles between the floor and the bottom of the case. If the case is fitted with leveling screws, they are adjusted to bring the case to a level position. When the case units are plumb and level, they are ready for fastening in place. They may be fastened to the wall with screws or nails placed through the fastening strip at the back of the cabinet (see Fig. 10-6).

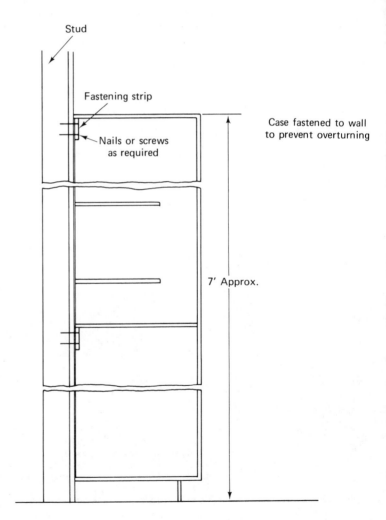

FIG. 10-6. Fastening Display Case to Wall.

Sometimes it is desirable to fasten the case to the floor. This is best done with angle brackets, which are attached to the cabinet first and then fastened to the floor. If the design of the cabinet will not allow the brackets attached to the cabinet to be fastened to the floor, it is necessary to set the cabinet in place and mark the location of the bracket on both the cabinet and the floor.

After marking the locations of the brackets, the display case is moved aside and the brackets are attached to the floor. The display case is then put back in position and fastened to the brackets (see Fig. 10-7).

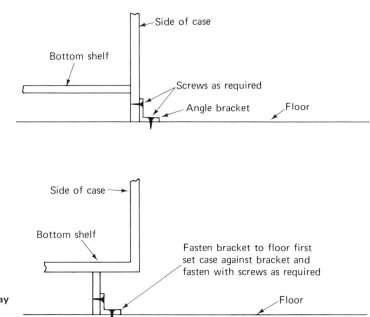

FIG. 10-7. Fastening Display Case to Floor.

FIXTURE WORK

Fixture work includes setting many types of display cases, counters, and decorative display platforms. These fixtures are almost always completely finished units that need only be located in the proper position, assembled, and fastened in place.

Counter High Display Cases

Fixtures are located according to the dimensions on the floor plan. Special care should be exercised in marking the locations of fixtures so that the required aisle width and work space behind the counters are maintained. Marks are placed on the floor that indicate the corners of the fixtures; and as the fixture locations are marked out in an area, the dimensions are rechecked.

Following a recheck of the layout, the fixtures may be set in place on the layout marks. Some of these units will be fastened to the floor by various means whereas others will be set in place and not fastened at all.

Some counters are made with a removable base unit (see Fig. 10-8). This base unit is located according to the layout marks and leveled in place. Usually this leveling is done with a 48 inch spirit level and wood shingles that are used to shim the base unit to bring it to a level position. The unit may be fastened to the floor with long screws running through holes drilled in the base, or it may be fastened with metal angle

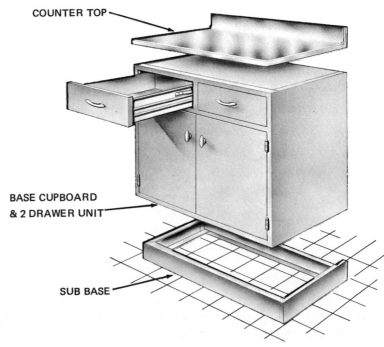

FIG. 10-8. Counter with Removal Base Unit.

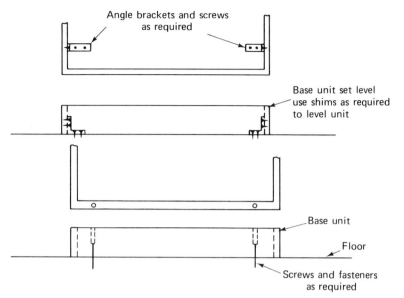

FIG. 10-9. Fastening Base Unit.

brackets (see Fig. 10-9). After the base is secured to the floor, the counter unit is set on the base. It may be fastened to the base with screws, or if it is a large heavy unit, not fastened at all.

Counters that are not fastened to the floor are leveled by means of adjustment screws in the legs or base of the unit (see Fig. 10-10). The adjustment screws are often hidden by some type of metal or rubber trim. This trim is raised out of the way to expose the adjusting screw and pulled down to the floor after the adjustment is completed.

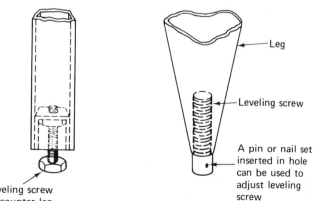

FIG. 10-10. Leveling Free Standing Counters.

Cornice Height Display Cases

High display cases, like other types, are located after studying the plans in the same manner as for counter high cases. They are set in place according to the layout marks and leveled (see Fig. 10-11). If they are high, narrow units, they must be securely fastened to the floor as indicated by the details furnished with the plans. If they are not anchored securely, there is danger that the unit may be tipped.

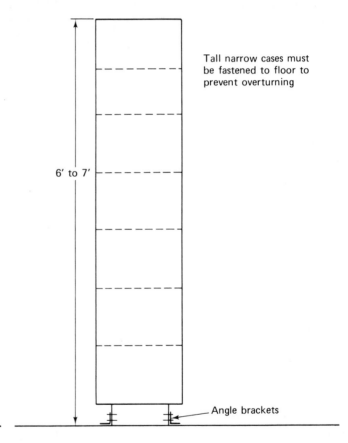

FIG. 10-11. High Display Case.

Display units that are high and wide may not require fastening to make them stable. This is generally true if two or more units meet to form one or more corners. They are fastened together at the corners, however, to prevent the unit from being overturned.

Installing Shelves, Doors, and Drawers. Nearly all display fixtures require the installation of shelves, doors, and drawers. All of these units are pre-fitted at the factory and only require to be put in place.

To install shelves in display cases, support brackets must be placed at the proper level before setting the shelves in place. These brackets are made in a variety of shapes (see Fig. 10-12) and are always furnished with the cabinets. They are simply set in the slot at the proper height and snapped into place.

FIG. 10-12. Shelf Support Brackets. (A) Shelf Standard with Supports in Place. (B) Supports.

Doors and drawers for fixtures are pre-fitted and usually pre-drilled to receive hardware. These units are usually set in place according to identification marks, but if there are no identifying marks they are sorted by trial and error until all the units are installed.

Display Platforms

Display platforms are platforms set 4 inches or more above the floor to set off the display from the floor (see Fig. 10-13). These platforms may be set along walls, at the end of counters, or in the center of a floor area.

Display platforms are set on a base that may be fastened to the floor or they may be free standing. If they are fastened to the floor, the procedure for installing Counter High Display Cases, as discussed earlier in this chapter, is followed.

CABINETS AND FIXTURE WORK 242

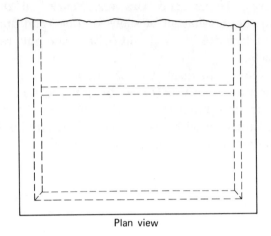

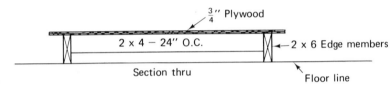

FIG. 10-13. Display Platform.

CABINETRY

Cabinets of all sizes and shapes are installed by carpenters in all types of commercial buildings. These cabinets may be made from wood and wood products, metal, plastic, or a combination of these materials. Although the size and shape of the cabinets may vary greatly, they may be divided into four categories for the purpose of outlining installation procedure. These are wardrobes, wall cabinets, base cabinets, and workbenches or work counters.

The first step in preparing for cabinet installation is to study the plans and determine the location of each unit. After this is done, the various areas can be prepared for the cabinets. The required preparation varies with the job and may require placing furring strips or temporarily removing pipes and electric fixtures.

Wardrobe Installation

A wardrobe is a tall cabinet approximately 7 feet high. It may be a unit used strictly for hanging clothes, or a combination wardrobe and storage cabinet, or a complete storage cabinet. Regardless of its function, the installation procedure will be the same.

These high units are sometimes set at the end of a row of wall or base cabinets, and it may be necessary to set the wall and base units first in order to determine the exact location for the tall cabinet unit.

After determining the location of the cabinet, it is moved into place, leveled and plumbed. To level and plumb the cabinet, cedar shingles or other shims are driven between the floor and the cabinet unit (see Fig. 10-14). The cabinet is fastened to the wall with nails or screws placed through the back of the cabinet into the wood or metal framework. If fastening to hollow walls, some type of hollow wall fastener such as toggle bolts or hollow wall anchors must be used.

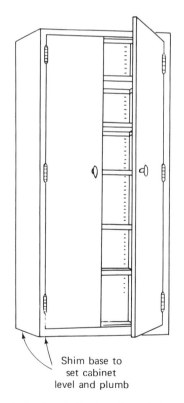

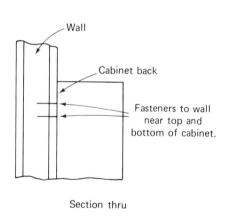

FIG. 10-14. Setting Tall Cabinets.

Wall Cabinets

Wall cabinets are units 30 inches to 36 inches high that are hung on the wall with the top edge approximately 84 inches above the floor. To determine the location of wall cabinets and the required height above the floor, the floor plans and elevations must be consulted. Detail drawings must be checked to determine the methods of fastening the cabinets to the wall.

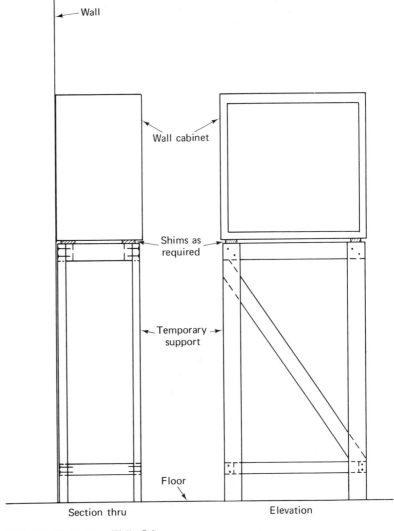

FIG. 10-15. Setting Wall Cabinets.

After determining the locations of the cabinets, they are moved into position and held in place until they can be fastened to the wall. It is usually difficult to hold these units by hand while they are being fastened in place, so temporary props or platforms should be used. Temporary platforms are especially useful if the installation involves a number of cabinets of the same height.

The platform is built slightly lower than required height so that the cabinet can be slid under a cornice (see Fig. 10-15). The cabinet is raised to final elevation with shims. These shims are removed after the cabinet is fastened in place, and the temporary support is moved to the next location. The temporary support must be provided with some type of shim arrangement, or it will be impossible to remove the support if it cannot be freed from the cabinet.

Wall cabinets are attached to the wall with nails or screws placed through the reinforcing or fastening strip at the back of the cabinet (see Fig. 10-16). Care should be taken to use a sufficient number of fasteners as specified in the plans. These fasteners should anchor securely. If they miss the framework or do not hold for some other reason, they should be removed and additional fasteners used to secure the cabinet. Anchoring the cabinet securely is important, because they generally hold a heavy load of equipment or supplies and could be pulled down unless fastened securely.

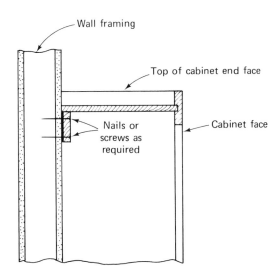

FIG. 10-16. Fastening Wall Cabinets.

CABINETS AND FIXTURE WORK

Base Cabinets

Base cabinets, as their name indicates, are lower cabinet units set on the floor at the base of the wall. The location of these cabinets is determined by studying the floor plan and elevations. Wood cabinets are set into place according to the floor plans, leveled with the aid of shingles, and fastened to the wall by means of nails or screws placed in the fastening strip at the top of the cabinet (see Fig. 10-17). The cabinet may also be nailed to the floor. If it is necessary to fasten to a concrete floor, angle brackets are fastened to the floor with some type of masonry anchor and screw. The angle bracket is then fastened to the cabinet with wood screws.

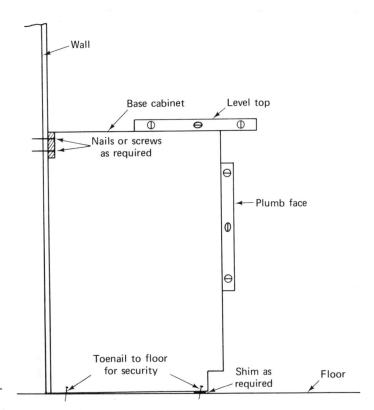

FIG. 10-17. Setting and Fastening Wood Base Cabinets.

Metal base cabinets are sometimes made to rest on either a metal sub base or a masonry base (see Fig. 10-18). The cabinets are set in place on the base and fastened to the floor and wall as required to make the unit stationary.

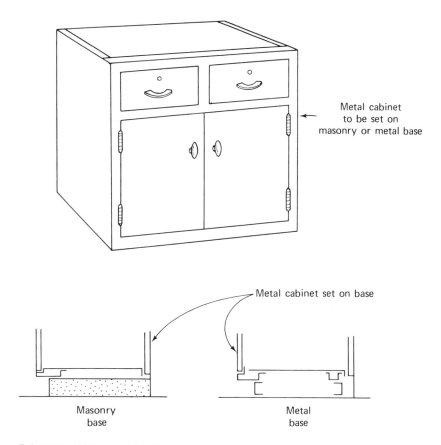

FIG. 10-18. Metal Base Cabinets.

Counter tops of various materials are installed after the base units are in place. The counter tops may be made of stainless steel, plastic laminates on plywood or hardboard backing, or other materials. These counter units are usually installed by tradesmen specializing in counter installation.

Workbenches

Workbenches are fastened to the floor by means of angle brackets attached to the legs or sides of the workbench. After the workbench is properly located, the position of screw holes in each bracket is marked on the floor. The bench is moved aside and holes are drilled in the floor for the screws or screw anchors that will be used. After the screw anchors have been

CABINETS AND FIXTURE WORK

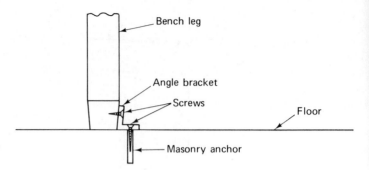

FIG. 10-19. Setting Work Benches.

placed, the workbench is relocated over the anchors and fastened in place (see Fig. 10-19).

REVIEW QUESTIONS

1. What are some of the various types of finish work done by carpenters on commercial jobs?
2. How are decorative cornices located and fastened in place?
3. How are inside corners fitted?
4. How are outside corners fitted?
5. How are cornice height display cases installed?
6. How are display cases fastened to the floor? The wall?
7. How are counter high display cases located?
8. How are counter high display cases fastened in place?
9. How are display platforms located and fastened in place?
10. Outline a procedure for installing wardrobe cabinets.
11. Outline a procedure for installing wall cabinets.
12. Outline a procedure for installing base cabinets.
13. Outline a procedure for installing workbenches.

RECOMMENDED PRACTICE FOR CONCRETE FORMWORK

APPENDIX A

The following pages contain the first three chapters of *Recommended Practice for Concrete Formwork* prepared by the American Concrete Institute, Committee 347. They are provided here to give the carpenter an opportunity to study the ACI recommendations that affect the majority of the formwork he will build.

Recommended Practice for Concrete Formwork (ACI 347-68)*

NEED FOR STANDARD

Since the cost of the formwork for a concrete structure may be 35 to 60 percent of the total cost of concrete work in the project, its design and construction demand sound judgment and planning to achieve adequate forms that are both economical and safe. The engineer or architect responsible for the successful completion of any concrete structure usually will include in his specifications provisons for stripping time, reshoring of concrete in place, inspection, and approval of formwork procedure which could affect the strength and appearance of the completed structure.

Neat, well-built, heavily braced forms may still fail due to inadequately tied corners or insufficient provision against uplift. Form failures have occurred when shoring has been improperly spliced, inadequately cross braced, or was otherwise inadequate to resist all possible stresses. Shores supported on previously completed floors usually can be assumed to have equal unit bearing. However, shores supporting forms for the first level above ground often are supported on "mudsills" and may not have uniform bearing. This condition may occur when mudsills rest on soft ground, on backfill recently placed and perhaps softened by surface water, or on frozen ground which may thaw out in numerous ways.

Unequal settlement of mudsills seriously changes shore reactions, and may cause serious overloading of shores which do not settle as much as others.

Proper engineering design of formwork often saves contractors more than the saving from the use of poorly designed forms. Formwork is generally more economically constructed on the ground and under the contractor's yard inspection than in the air under field conditions. A carpenter not experienced in formwork may find it difficult to conceive of forms as pressure vessels which need to be adequately tied together, braced, and anchored to resist uplift and to be capable of resisting forces in several directions. Proper detailed instruction is therefore necessary.

If working drawings are made for formwork, the necessary detailed study of the contract drawings may uncover omissions and dimensional errors. Field work is expedited and the structural engineer can see how his design is being interpreted by the contractor. Other benefits may be shorter job duration, avoidance of delays in field operations, more efficient re-use of forms, and better utilization of material and men. There should be "notes to the erector" on such drawings to eliminate need for referring to specific field customs.

ACI STANDARD

*Courtesy of American Concrete Institute.

Attention is called to the legal implications of specifying in any set of contract documents, including the plans and specifications, both the method by which work is to be performed and the results to be accomplished.

Some contractors feel that "end result" or performance specifications are sufficient for a contract involving concrete construction. This simple approach may be best when bidding is restricted to highly experienced contractors and for special projects which require competitive formwork solutions. In such special instances specifications must allow freedom to improvise and to apply a choice of equipment and materials.

For the majority of formwork jobs, however, certain rules and regulations are needed such as minimum design loadings, safe unit design stresses, and tolerances. Such general minimum requirements should be clearly stated to assure the owner and his engineer or architect that formwork will provide adequate support during the placement of concrete and until it has gained sufficient strength to permit removal of forms. Adequately written and enforced specifications and contractor observance of them should lessen the need for legislation or for provisions in building construction safety codes to govern design of formwork, its erection, and its removal.

It is the conclusion of the committee that the layout and design of the formwork, as well as its construction, must be the responsibility of the contractor. This approach gives him the necessary freedom to use his skill and knowledge to produce an economical finished structure.

ENGINEER-ARCHITECT* SPECIFICATIONS

GENERAL

For any concrete structure, the specification for formwork, if written by the engineer or architect, will have much to do with the over-all economy and quality of finished work. *Such a specification must be individualized for the particular job, must indicate to the contractor exactly what will be expected from him, and must be so written as to result in economy and safety.*

A well-written formwork specification tends to equalize bids for the work, provided each bidder knows that full compliance will be required of the successful one. Unnecessarily exacting requirements may make bidders question the specification as a whole and may render it virtually impossible for them to understand exactly what is expected. They may be overly cautious and overbid or not cautious enough and underbid.

A well-prepared formwork specification is of value not only to the owner and the contractor, but also to the field representative of the engineer-architect and to the subcontractors for other trades.

This proposed standard presents recommended practices for formwork and suggests criteria for assuring proper performance.

Some requirements are so written as to allow the contractor discretion where quality of finished concrete work would not be impaired by the use of alternate materials and methods. The engineer-architect may exclude, call special attention to, or strengthen, or make more lenient any requirement to best fit the needs of his particular project.

Consideration of the applicable general requirements suggested herein will not, however, be sufficient to make the specification complete. To it there must be added requirements for actual materials, finishes, and other items peculiar to and necessary for the individual structure. Much helpful and detailed information is given in *Formwork for Concrete,* ACI Special Publication No. 4.

Formwork materials and accessories

If the particular design or desired finish requires special attention, the engineer or architect may specify in his contract plans and specifications, formwork materials and such other features he feels necessary to attain his objectives. If he does not call for specific materials or accessories the contractor will be free to use materials of his choice as long as they meet design requirements.

When structural design is predicated on the use of a commercially available form unit in standard sizes such as one-way and two-way joist systems, plans and specifications should be drawn up realistically to make use of available shapes and sizes. Some latitude must be permitted for connections of form units to other framing or centering to reflect the tolerances and normal installation practices of the form type contemplated.

Finish of exposed concrete

Finish requirements for concrete surfaces should be described in *measurable* terms as precisely as practicable.

Design, inspection, and approval of formwork

The following items should be clarified in the engineer-architect specifications and drawings:

(a) By whom formwork will be designed.

---
*The terms *engineer-architect* and *architect-engineer* are used interchangeably throughout the report to designate: the architect, the engineer, the architectural firm, the engineering firm, the architectural and engineering firm, or other agency issuing project drawings and specifications and/or administering the work under project specifications and drawings.

(b) By whom, when, and for what features formwork will be inspected.

(c) What approvals will be required for for formwork drawings; for the forms before concreting and during concreting; and for form removal and reshoring; and who will give such approvals.

ACHIEVING ECONOMY IN FORMWORK

The suggestions which follow are typical of those received from contractors as to how the engineer-architect can plan his work to reduce formwork costs.

The engineer-architect should consider not only the type of formwork necessary for the quality of construction desired, but also how economy of construction can be achieved. Since the cost of formwork is a significant part of the over-all concrete cost in a structure, a saving in this item alone can result in substantial reductions in the total cost. The following points should be considered and made use of insofar as possible:

1. If column and floor form dimensions can be the same size from the lowest deck to roof, beam forms as well as column forms can be re-used from floor to floor without alteration.

2. If spacing of columns and story heights are made uniform throughout the building insofar as possible, formwork will be simplified and its prefabrication more economically feasible.

3. If interior columns are the same width as or smaller than the girders they support, the column form becomes a simple rectangular or square box without cutouts, and the slab form does not have to be cut out at each corner of the column.

4. If all beams are made one depth (beams framing into beams as well as beams framing into columns), the supporting structure for the beam forms can be carried on a level platform supported on shores.

If the engineer-architect makes widths and depths the same for beams and joists, and considers the available sizes of dressed lumber, plywood, and the various ready-made formwork components when determining the sizes of structural members, savings in labor time in cutting, measuring, and leveling the work will be achieved.

5. Where commercially available forming systems such as one-way or two-way joist systems are used, design should be based on the use of one standard size range wherever possible.

6. Structural design should be prepared simultaneously with the architectural design, so that dimensions can be better coordinated. Room sizes can often be varied a few inches to accommodate the structural design.

7. The engineer-architect should be responsible for coordinating the requirements of other trades with their effect on the concrete formwork. Architectural features, depressions, and openings for mechanical or electrical work should be coordinated with the structural system for maximum economy to the over-all job, and variations in the structural system caused by such items should be shown on the structural drawings. Wherever possible depressions in the tops of slabs should be made without a corresponding break in elevations of the soffits of slabs, beams, or joists.

CHAPTER 1–DESIGN

1.1—General

1.1.1—Planning—Any form regardless of size should be planned in every particular prior to its construction. The thoroughness of planning required will depend on the size, complexity, and importance (considering re-uses) of the form. In any case, all of the applicable details listed in Section 1.4.2.3. should be included in the planning.

1.1.2—Design and erection—Formwork should be designed, erected, supported, braced, and maintained so that it will safely support all vertical and lateral loads that might be applied until such loads can be supported by the concrete structure. Vertical and lateral loads must be carried to the ground by the formwork system and by the in-place construction that has attained adequate strength for that purpose. Formwork should also be constructed so that concrete slabs, walls, and other members will be of correct size in dimensions, shape, alignment, elevation, and position.

1.2—Loads

1.2.1—Vertical loads—Vertical loads consist of a dead load plus an allowance for live load. The weight of formwork together with the weight of freshly placed concrete is dead load. The live load consists of the weight of workmen, equipment, runways, and impact, and should be taken as not less than 50 psf of horizontal projection.

1.2.2—Maximum lateral pressure of concrete—Forms, ties, and bracing should be designed for a lateral pressure of fresh concrete as follows:

(a) For ordinary work with normal internal vibration, in columns,

$$p = 150 + \frac{9000R}{T} \quad \text{(maximum 3000 psf or 150}h\text{, whichever is least)}$$

in walls, with rate of placement not exceeding 7 ft per hr

$$p = 150 + \frac{9000R}{T}$$ (maximum 2000 psf or $150h$, whichever is least)

and in walls, with rate of placement greater than 7 ft per hr

$$p = 150 + \frac{43{,}400}{T} + \frac{2800R}{T}$$

(maximum 2000 psf or $150h$, whichever is least)

where

p = lateral pressure, psf
R = rate of placement, ft per hr
T = temperature of concrete in the forms, deg F
h = height of fresh concrete above point considered, ft

(b) For concretes weighing other than 150 lb per cu ft; containing pozzolanic additions or cements other than Type I; having slumps greater than 4 in.; or consolidated by revibration or external vibration of forms, appropriate adjustment for increased lateral pressure should be made.*

(c) Where retarding admixtures are employed under hot weather conditions an effective value of temperature less than that of the concrete in the forms should be used in the above formula.

If retarding admixtures are used in cold weather the lateral pressure should be assumed as that exerted by a fluid with weight equal to that of the concrete mix.

1.2.3—*Lateral loads*—Braces and shores should be designed to resist all foreseeable lateral loads such as wind, cable tensions, inclined supports, dumping of concrete, and starting and stopping of equipment. In no case should the assumed value of lateral load due to wind, dumping of concrete, and equipment acting in any direction at each floor line be less than 100 lb per lineal ft of floor edge or 2 percent of total dead load of the floor, whichever is greater. Wall forms should be designed for a minimum wind load of 10 psf, and bracing for wall forms should be designed for a lateral load at least 100 lb per lineal ft of wall, applied at the top. Walls of unusual height and exposure should be given special consideration.

Wind loads on enclosures or other wind breaks attached to the forms should be considered, as well as those applied to the forms themselves.

1.2.4—*Special loads*—The formwork should be designed for any special conditions of construction likely to occur, such as unsymmetrical placement of concrete, impact of machine-delivered concrete, uplift, and concentrated loads of reinforcement and storage of construction materials. Form designers should be alert to provide for special loading conditions such as walls constructed over spans of slabs or beams which exert a different loading pattern before hardening of concrete than that for which the supporting structure is designed.

Imposition of any construction loads on the partially completed structure should not be allowed except with the approval of the engineer-architect. See Section 2.8 for special conditions pertaining to multistory work.

1.3—Design considerations

1.3.1—*Unit stresses*—Unit stresses for use in the design of formwork, exclusive of accessories, are given in the applicable codes or specifications listed in Chapter 3, Materials for Formwork. When fabricated formwork, shoring, or scaffolding units are used, manufacturers' recommendations for allowable loads may be followed if supported by test reports or successful experience records; for materials which will experience substantial re-use, reduced values may be required.

For forms of a temporary nature with limited re-use, allowable stresses should be those specified in the appropriate design codes or specifications for temporary structures or for temporary loads on permanent structures.

Where there will be a considerable number of form re-uses or where forms to be used many times are fabricated from material such as steel, aluminum, or magnesium, it is recommended that the formwork be designed as a permanent structure carrying permanent loads.

In the design of formwork accessories such as form ties, form anchors, and form hangers the following minimum safety factors based on the ultimate strength of the accessory are recommended except that yield point must not be exceeded (see Table 1.3.1).

1.3.2—*Analysis*—A design analysis should be made for all formwork. Safety against buckling of any member should be investigated in all cases.

1.3.3—*Shores*— Shores are defined as vertical or inclined falsework support. When patented shores, patented splices in shoring, or patented methods of shoring are used, manufacturers' recommendations as to load-carrying capacities may be followed but only if supported by test reports by a qualified and recognized testing laboratory; the designer must carefully follow the manufacturer's recommendations as to bracing and working loads for unsupported shore lengths.

---
*Detailed information on the effect of these special conditions is available in the report "Pressures on Formwork," ACI Committee 622, ACI JOURNAL, *Proceedings* V. 55, No. 2, Aug. 1958, pp. 173-190, and in the discussion and committee's closure for this report, ACI JOURNAL, *Proceedings* V. 55, No. 12, June 1959, pp. 1335-1348.

Field constructed lap splices must not be used more often than for alternate shores under slabs, or for every third shore under beams, and shores should not be spliced more than once unless diagonal and two-way lateral bracing is provided at every splice point. Such spliced shores should be distributed as uniformly as possible throughout the work. To avoid buckling, splices should not be located near the midheight of the shores nor midway between points of lateral support.

Splices must be designed against buckling and bending as for any other structural compression member. The minimum length of splice material for timber shores should be 2 ft. Shores made of round timbers should have three splice pieces at each splice, and those of square timbers should have four splice pieces at each splice.

Splicing material should be not less than 2-in. (nominal) lumber or ⅝-in. plywood and no less than the width of the material being spliced.

1.3.4—*Diagonal bracing*—The formwork system must be designed to transfer all lateral loads to the ground or to completed construction in such a manner as to insure safety at all times. Diagonal bracing must be provided in vertical and horizontal planes where required to provide stiffness and to prevent buckling of individual members. Where the only bracing requirement is to prevent buckling of individual members, lateral bracing should be provided in whatever directions are necessary to produce the correct l/r ratio for the load supported. Such a laterally braced system should be anchored in such a manner as to insure stability of the total system.

1.3.5—*Foundations for formwork*—Proper foundations on ground, mudsills, spread footings, or pile footings must be provided. If soil under mudsills is or may become incapable of supporting superimposed loads without appreciable settlement, it should be stabilized with cement or lean concrete, or by other adequate methods. Mudsills should never be placed on frozen ground.

1.3.6—*Settlement*—Falsework should be so constructed that vertical adjustments can be made to compensate for take-up and settlements, and so that settlements under full load will be a minimum consistent with economy. Where wood timbers are used, the number of horizontal joints and particularly the number of joints where end grain bears on side grain should be kept at a practical minimum. The unit compressive stress across the grain should not exceed that recommended in Section 3.2. The probable settlement of the falsework, exclusive of foundations, may be approximated as follows: (1) by computing the columnar shortening c in in., by the formula $c = 12SL/E$, in which S is the unit compressive stress per sq in., L is the length of column in ft, and E is the modulus of elasticity; (2) by allowing an assumed value of settlement for each horizontal joint for "taking up" and an additional value at each joint where end grain bears on side grain for "biting" of the end grain into the side grain. For normal carpentry work each assumed value may be taken as 1/16 in.; where particular care is taken, each may be taken as 1/32 in. The total estimated settlement should approximate the sum of (1) and (2).

Wedges may be used at the top or bottom of shores, but not at both ends, to facilitate vertical adjustment, to correct uneven settlements, or to facilitate dismantling of the formwork.

Screw jacks for pipe shores or scaffold-type shoring may be used at both top and bottom so

TABLE 1.3.1—DESIGN CAPACITIES OF FORMWORK ACCESSORIES*

Accessory	Safety factor	Type of construction
Form tie	1.5	Light formwork; or ordinary single lifts at grade and 16 ft or less above grade
	2.0	Heavy formwork; all formwork more than 16 ft above grade or unusually hazardous
Form anchor	1.5	Light form panel anchorage only; no hazard to life involved in failure
	2.0	Heavy forms—failure would endanger life—supporting form weight and concrete pressures only
	3.0	Falsework supporting weight of forms, concrete, working loads, and impact
Form hangers	1.5	Light formwork. Design load including total weight of forms and concrete, with 50 psf minimum live load is less than 150 psf
	2.0	Heavy formwork; form plus concrete weight 100 psf or more; unusually hazardous work
Lifting inserts	2.0	Tilt-up panels
	3.0	Precast panels
Expendable strand deflection devices†	2.0	Pretensioned concrete members
Re-usable strand deflection devices†	3.0	Pretensioned concrete members

*Design capacities guaranteed by manufacturers may be used in lieu of tests for ultimate strength.
†These safety factors also apply to pieces of prestressing strand which are used as part of the deflection device.

long as they are secured by the shore or scaffold leg against loosening or falling out. A minimum of 8 in. embedment in the pipe leg or sleeve should be required, or manufacturers' recommendations based on test results may be followed.

1.4—Drawings

1.4.1—*Discussion*—Should the engineer-architect wish formwork drawings submitted for his approval, he should so state in the specifications. In any case, the contract plans and specifications should include and cover all points necessary to the contractor for his formwork design and in the preparation of his formwork drawings, such as:

(a) Number, location, and details of all construction joints, contraction joints, and expansion joints that will be required or permitted for the particular job or parts of it.

(b) Sequence of placement, if critical, should be indicated.

(c) Locations of and details for architectural concrete. When architectural details are to be cast into structural concrete, they should be so indicated or referenced on the structural drawings as they may play a key role in the structural design of the form.

(d) Intermediate supports under permanent forms (such as metal deck used for forms, and permanent forms of other materials), supports required by the structural engineer's design for composite action, and any other special supports.

(e) The location and order of erection and removal of shores for composite construction.

(f) Special provisions essential for formwork of the special structures and special construction methods such as shells and folded plates.

The basic geometry of such structures, as well as the required camber, must be given in sufficient detail to permit the contractor to construct the form. Camber should be stipulated for measurement after initial set and before decentering.

(g) Special requirements for post-tensioned concrete members. The effect of load transfer during tensioning of post-tensioned members may be critical, and the contractor should be advised of any special provisions that must be made in the formwork for this condition.

(h) If camber is desired for slab soffits or structural members to compensate for elastic deflection and/or deflection due to creep of the concrete, the contract drawings must so indicate and state the amounts. Measurement of camber attained should be made *after* initial set and *before* decentering.

The question of allowable camber for precast members, especially prestressed precast members, is an important one. The engineer-architect must specify the amount of camber for any member and the difference in camber between adjacent members which will be acceptable in finished work. Any devices (such as clips to be embedded in edges of members and to be field welded together) desired to reduce or control differential camber should be indicated on the design drawings.

(i) Where chamfers are required on beam soffits or column corners, they should be specified.

(j) Plans and specifications of the engineer-architect must cover in detail any requirements for inserts, waterstops, built-in frames for openings, holes through concrete, and similar requirements were work of other trades will be attached to or supported by formwork.

(k) Where architectural features, embedded items or the work of other trades will change the location of structural members such as joists in one-way or two-way joist systems, such changes or conditions should be indicated on the structural drawings.

(l) The ACI Building Code requirement that structural drawings shall show the live load used in the design should always be followed.

1.4.2—*Recommendations*

1.4.2.1—*General*—Before constructing forms, the contractor, if required, will submit detailed drawings of proposed formwork for approval by the engineer. If such drawings are not in conformity with contract documents as determined by the engineer-architect, the contractor will make such changes as may be required prior to start of work.

1.4.2.2—*Design assumptions*—All major design values and loading conditions should be shown on formwork drawings. These include assumed values of live load, rate of placement, temperature of concrete, height of drop, weight of moving equipment which may be operated on formwork, foundation pressures, design stresses, camber diagrams, and other pertinent information, if applicable.

1.4.2.3—*Items included*—In addition to specifying types of materials, sizes, lengths, and connection details, formwork drawings should provide for applicable details such as:

1. Sequence of removal of forms and shores (when this is critical for placing loads on new concrete)

2. Design allowance for construction loads on new slabs should be shown when such

allowance will affect the development of shoring and/or reshoring schemes (see Section 2.8)

3. Anchors, form ties, shores, and braces
4. Field adjustment of the form during placing of concrete
5. Waterstops, keyways, and inserts
6. Working scaffolds and runways
7. Weepholes or vibrator holes where required
8. Screeds and grade strips
9. Crush plates or wrecking plates where stripping may damage concrete
10. Removal of spreaders or temporary blocking
11. Cleanout holes
12. Construction joints, control joints, and expansion joints to conform to design drawings [ACI Building Code (318-63) Section 704]
13 Sequence of concrete placements and minimum elapsed time between adjacent placements
14. Chamfer strips or grade strips for exposed corners and construction joints
15. Camber [see Section 2.6.1 (d)]
16. Mudsills or other foundation provisions for falsework
17. Special provisions such as protection from ice and debris at stream crossings, fire, and safety
18. Formwork coatings
19. Notes to formwork erector for conduits and pipes embedded in concrete according to ACI Building Code (318-63), Section 703

1.5—Approval by the engineer or architect

Although the safety of formwork is the responsibility of the contractor, the engineer or architect may, under certain circumstances, wish to review or approve the formwork design. If so, the engineer-architect will include provisions for such review or approval in his specifications.

Approval might be required for unusually complicated structures, for structures whose designs were predicated on a particular method of construction, for structures in which the forms impart a desired architectural finish, for certain post-tensioned structures, for folded plates, for thin shells, and for long-span roof structures.

The engineer-architect's approval of the drawings as submitted or as corrected in no way relievese the contractor of his responsibility for adequately constructing and maintaining the forms so that they will function properly. Such approval indicates that the assumed design loadings in combination with design stresses shown; proposed construction methods, rates, equipment, and sequence; the proposed form materials; and the overall scheme of formwork are deemed capable of producing the desired concrete in an approved manner.

CHAPTER 2—CONSTRUCTION

2.1—Safety precautions

In addition to the very real moral and legal responsibility to maintain safe conditions for workmen and the public, safe construction is in the final analysis more economical, irrespective of any short-term cost savings from cutting corners on safety provisions. Attention to safety is particularly significant in form construction as these structures support the concrete during its plastic state and as it is developing its strength, at which time it is unstable. Following the design criteria contained in this standard is essential to assuring safe performance of the forms. All structural members and their connections should be carefully planned so that a sound determination of loads thereon may be accurately made and allowable stresses calculated.

In addition to the adequacy of design of forms, multistory work requires further consideration of the shoring of newly completed slabs below which support the weight of fresh concrete as well as other construction loads (see Section 2.8).

Many form failures can be attributed to some human error or omission rather than basic inadequacy in design. Careful direction and inspection of formwork erection by qualified members of the contractor's organization can prevent many accidents.

Some common construction deficiencies leading to form failures are:

(a) Inadequate diagonal bracing of shores

(b) Inadequate lateral and diagonal bracing and poor splicing of "double-tier shores" or "multiple-story shores"

(c) Failure to control rate of placing concrete vertically without regard to drop in temperature

(d) Failure to regulate properly the rate and sequence of placing concrete horizontally to avoid unbalanced loadings on the formwork

(e) Unstable soil under mudsills

(f) Failure to inspect formwork during and after concrete placement to detect abnormal deflections or other signs of imminent failure which could be corrected

(g) Insufficient nailing

(h) Failure to provide for lateral pressures on formwork

(i) Shoring not plumb and thus inducing lateral loading as well as reducing vertical load capacity

(j) Locking devices on metal shoring not locked, inoperative, or missing

(k) Vibration from adjacent moving loads or load carriers

(l) Inadequately tightened or secured form ties or wedges

(m) Form damage in excavation by reason of embankment failure

(n) Loosening of reshores under floors below

(o) Premature removal of supports, especially under cantilevered sections

Some common design deficiencies leading to failure are:

(a) Lack of proper field inspection by qualified persons to see that form design has been properly interpreted by form builders

(b) Lack of allowance in design for such special loadings as wind, power buggies, placing equipment

(c) Inadequate reshoring

(d) Improperly stressed reshoring

(e) Improper positioning of shores from floor to floor which creates reverse bending in slabs which are not designed for such stresses

(f) Inadequate provisions to prevent rotation of beam forms where slabs frame into them on only one side

(g) Inadequate anchorage against uplift due to battered form faces

(h) Insufficient allowance for eccentric loading due to placement sequences

Specific safety provisions which should be considered are:

(a) Erection of safety signs and barricades to keep unauthorized personnel clear of areas in which erection or stripping is under way

(b) Providing form watchers during concrete placement wherever there is danger to life or property from forms failing or distorting during placement

(c) Furnishing extra shores or other material and equipment that might be needed in an emergency by form watchers

(d) Incorporation of scaffolds, guard rails, etc., into form design where feasible

2.2—General practices

Construction procedures must be planned in advance to insure the unqualified safety of personnel engaged in formwork and concrete placement and the integrity of the finished structure.

Forms should be inspected and checked before the reinforcing steel is placed to insure that the concrete will have the dimensions and be in the location shown on the drawings.

Forms should be sufficiently tight to prevent loss of mortar from the concrete.

Forms should be thoroughly cleaned of all dirt, mortar, and foreign matter and coated with a release agent before each use. Where the bottom of the form is inaccessible from within, access panels should be provided to permit thorough removal of extraneous material before placing concrete. If surface appearance is important, forms should not be re-used after damage from previous use has reached the stage of possible impairment to concrete surfaces.

Bulkheads for control joints or construction joints should preferably be made by splitting along the lines of reinforcement passing through the bulkhead so that each portion may be positioned and removed separately without applying undue pressure on the reinforcing rods which could cause spalling or cracking of the concrete. When required on the engineer-architect's drawings, beveled inserts at control joints must be left undisturbed when forms are stripped, and removed only after the concrete has been sufficiently cured and dried out. Wood strips inserted for architectural treatment should be kerfed to permit swelling without pressure on the concrete.

Sloped surfaces in excess of 35 deg from the horizontal (1.5 horizontal to 1 vertical) should be provided with a top form to hold the shape of the concrete during placement, unless there is a continuous mat of bars or mesh sufficient to keep concrete in place.

Loading of new slabs should be avoided in the first few days after placement. Loads such as aggregate, timber, boards, reinforcing steel, or support devices, must not be thrown on new construction, nor be allowed to pile up in quantity.

Building materials must not be thrown or piled on the formwork in such manner as to damage or overload it.

2.3—Workmanship

To insure good workmanship the following points warrant careful attention:

(a) Proper splices of studs, wales, or shores

(b) Staggering of joints or splices in sheathing, plywood panels, and bracing

(c) Proper seating of shores

(d) Proper number and location of form ties or clamps

(e) Proper tightening of form ties or clamps

(f) Adequate bearing under mudsills; *in no case should mudsills or spread footings rest on frozen ground*

(g) Connection of shores to joists, stringers, or wales must be adequate to resist uplifts or torsion at joints

(h) Form coatings must be applied before placing of reinforcing steel and must not be used in such quantities as to run onto bars or concrete construction joints

(i) Details of control joints, construction joints, and expansion joints.

2.4—Suggested tolerances

Tolerance is a specified permissible variation from lines, grades, or dimensions given in contract drawings.

Tolerances should be specified by the engineer-architect so that the contractor will know precisely what is required and can design and maintain his forms accordingly. The suggested tolerances herein are similar to those specified on important work or major structures by many public agencies and private firms.* In specifying these tolerances or some modifications of them, it should be remembered that specifying tolerances more exacting than needed may increase construction costs.

Contractors are expected, and should be required, to establish and maintain in an undisturbed condition until final completion and acceptance of a project, control points and bench marks adequate for their own use and for reference to establish tolerances. (This requirement may become even more important for the contractor's protection when tolerances are not specified or shown.) The engineer-architect should specify tolerances or require performance within generally accepted limits. Where a project involves particular features sensitive to the cumulative effect of generally accepted tolerances on individual portions, the engineer-architect should anticipate and provide for this effect by setting a cumulative tolerance. Where a particular situation involves several types of generally accepted tolerances, i.e., on form, on location of reinforcement, on fabrication of reinforcement, etc., which become mutually incompatible, the engineer-architect should anticipate the difficulty and specify special tolerances or indicate which controls.

The engineer-architect should be responsible for coordinating the tolerances for concrete work with the requirements of other trades whose work adjoins the concrete construction.

This section suggests tolerances that are consistent with modern construction practice, considering the effect that permissible deviations will have on the structural action or operational function of the structure. Surface defects such as "blow holes" and "honeycomb" concrete surfaces are defined as "finishes" and are to be distinguished from tolerances described herein.

Where tolerances are not stated in the specifications or drawings for any individual structure or feature thereof, permissible deviations from established lines, grades, and dimensions are suggested below. The contractor is expected to set and maintain concrete forms *so as to insure completed work within the tolerance limits.*

No tolerances specified for horizontal or vertical building lines or footings should be construed to permit encroachment beyond the legal boundaries.

2.4.1—*Tolerances for reinforced concrete buildings*†

1. *Variation from the plumb*
   (a) In the lines and surfaces of columns, piers, walls, and in arrises ¼ in. per 10 ft, but not more than 1 in.
   (b) For exposed corner columns control-joint grooves, and other conspicuous lines
   In any bay or 20 ft maximum .... ¼ in.
   In 40 ft or more .... ½ in.

2. *Variation from the level or from the grades indicated on the drawings*
   (a) In slab soffits,‡ ceilings, beam soffits, and in arrises
   In 10 ft .... ¼ in.
   In any bay or 20 ft maximum .... ⅜ in.
   In 40 ft or more .... ¾ in.
   (b) For exposed lintels, sills, parapets, horizontal grooves, and other conspicuous lines
   In any bay or 20 ft maximum .... ¼ in.
   In 40 ft or more .... ½ in.

3. *Variation of the linear building lines from established position in plan and related position of columns, wall and partitions*
   In any bay or 20 ft maximum .... ½ in.
   In 40 ft or more .... 1 in.

4. *Variation in the sizes and locations of sleeves, floor openings, and wall openings* .... ¼ in.

5. *Variation in cross-sectional dimensions of columns and beams and in the thickness of slabs and walls*
   Minus .... ¼ in.
   Plus .... ½ in.

---

*Designers employed by federal agencies required to follow Building Research Advisory Board recommendations are advised that the BRAB tolerances on formwork are often much more restrictive than those suggested herein.
†Variations from plumb and linear building lines on upper stories of high rise structures (above 100 ft high) are special cases which may require special tolerances.
‡Variations in slab soffits are to be measured *before* removal of supporting shores; the contractor is not responsible for variations due to deflection, except when the latter are corroboratory evidence of inferior concrete quality or curing, in which case only the *net* variation due to deflection can be considered.

6. *Footings*
   (a) Variation in dimensions in plan
       Minus .... ½ in.
       Plus .... 2 in.*
   (b) Misplacement or eccentricity
       2 percent of the footing width in the direction of misplacement but not more than .... 2 in.*
   (c) Reduction in thickness
       Minus .... 5 percent of specified thickness

7. *Variation in steps*
   (a) In a flight of stairs
       Rise .... ⅛ in.
       Tread .... ¼ in.
   (b) In consecutive steps
       Rise .... 1/16 in.
       Tread .... ⅛ in.

2.4.2.—*Tolerances for special structures*

1. *Concrete canal lining*
   (a) Departure from established alignment
       2 in. on tangents
       4 in. on curves
   (b) Departure from established profile grade .... 1 in.
   (c) Reduction in thickness of lining
       10 percent of specified thickness: *provided*, that average thickness is maintained as determined by daily batch volumes
   (d) Variation from specified width of section at any height
       ¼ of 1 percent plus 1 in.
   (e) Variation from established height of lining
       ½ of 1 percent plus 1 in.
   (f) Variations in surfaces
       Invert ¼ in. in 10 ft
       Side slopes ½ in. in 10 ft

2. *Monolithic siphons and culverts*
   (a) Departure from established alignment .... 1 in.
   (b) Departure from established profile grade .... 1 in.
   (c) Variation in thickness
       At any point: minus 2½ percent or ¼ in., whichever is greater
       At any point: plus 5 percent or ½ in., whichever is greater
   (d) Variation from inside dimensions
       ½ of 1 percent
   (e) Variations in surfaces:
       Inverts ¼ in. in 10 ft
       Side slopes ½ in. in 10 ft

3. *Bridges, checks, overchutes, drops, turnouts, inlets, chutes, and similar structures*
   (a) Departure from established alignment .... 1 in.
   (b) Departure from established grades 1 in.
   (c) Variation from the plumb or the specified batter in the lines and surfaces of columns, piers, walls, and in arrises
       Exposed, in 10 ft .... ½ in.
       Backfilled, in 10 ft .... 1 in.
   (d) Variation from the level or from the grades indicated on the drawings in slabs, beams, horizontal grooves, and railing offsets
       Exposed, in 10 ft .... ½ in.
       Backfilled, in 10 ft .... 1 in.
   (e) Variation in cross-sectional dimensions of columns, piers, slabs, walls, beams, and similar parts
       Minus .... ¼ in.
       Plus .... ½ in.
   (f) Variation in thickness of bridge slabs
       Minus .... ⅛ in.
       Plus .... ¼ in.
   (g) Footings: Same as for footings for buildings
   (h) Variation in the sizes and locations of slab and wall openings .... ½ in.
   (i) Sills and side walls for radial gates and similar watertight joints. Variation from the plumb or level
       Not greater than ⅛ in. in 10 ft.

4. *Tolerances in mass concrete structures*
   (a) All structures
       1. Variation of the constructed linear outline from established position in plan
          In 20 ft .... ½ in.
          In 40 ft .... ¾ in.
       2. Variations of dimensions to individual structure features from established positions
          In 80 ft or more .... 1¼ in.
          In buried construction
              Twice the above amounts
       3. Variation from the plumb, from the specified batter, or from the curved surfaces of all structures, including the lines and surfaces of columns, walls, piers, buttresses, arch sections, vertical joint grooves, and visible arrises
          In 10 ft .... ½ in.
          In 20 ft .... ¾ in.
          In 40 ft or more .... 1¼ in.
          In buried construction
              Twice the above amounts
       4. Variation from the level or from the grades indicated on the drawings in slabs, beams, soffits, horizontal joint grooves, and visible arrises
          In 10 ft .... ¼ in.
          In 30 ft or more .... ½ in.
          In buried construction
              Twice the above amounts

---

*Applies to concrete only, not to reinforcing bars or dowels.

5. Variation in cross-sectional dimensions of columns, beams, buttresses, piers, and similar members
   Minus .... ¼ in.
   Plus .... ½ in.

6. Variation in the thickness of slabs, walls, arch sections, and similar members
   Minus .... ¼ in.
   Plus .... ½ in.

(b) Footings for columns, piers, walls, buttresses, and similar members

1. Variation of dimensions in plan
   Minus .... ½ in.
   Plus .... 2 in.*

2. Misplacement or eccentricity
   2 percent of footing width in the direction of misplacement but not more than .... 2 in.*

3. Reduction in thickness
   5 percent of specified thickness

(c) Sills and side walls for radial gates and similar watertight joints

1. Variation from plumb and level
   Not greater than ⅛ in. in 10 ft

5. *Tolerances for concrete tunnel lining and cast-in-place conduits*

(a) Departure from established alignment or from established grade
   Free-flow tunnels and conduits   1 in.
   High velocity tunnels and conduits .... ½ in.
   Railroad tunnels .... 1 in.

(b) Variation in thickness at any point
   Tunnel lining .... minus 0
   Conduits .... minus 2½ percent or ¼ in, whichever is greater
   Conduits .... plus 5 percent or ½ in., whichever is greater

(c) Variations from inside dimensions
   ½ of 1 percent

2.5—Falsework and centering

2.5.1—*Falsework*—Falsework is the temporary structure erected to support work in the process of construction. It is composed of shores, formwork for the beams and/or slabs, and lateral bracing. The falsework must be supported on satisfactory foundations such as spread footings, mudsills, or piling as discussed in Section 1.3

Shoring resting on intermediate slabs or other construction already in place need not be located directly above shores or reshores below unless thickness of slab and the location of its reinforcement are inadequate to take the reversal of stresses. Where the latter conditions are questionable the shoring location should be approved by the engineer.

All members must be straight and true without twists or bends. Special attention should be given to beam and slab, or one-way and two-way joist construction to prevent local overloading when a heavily loaded shore rests on the thin slab.

Shores in multitier assemblies supporting forms for high stories must be set plumb and the separate parts of each shore located in a straight line over each other, with two-way lateral bracing at each splice in the shore unless the entire assembly is designed as a structural framework or truss. Particular care must also be taken to transfer the lateral loads to the ground or to completed construction of adequate strength (see Section 1.3.4).

Where a slab load is supported on one side of the beam only, edge beam forms should be carefully planned to prevent tipping of the beam due to unequal loading.

Shores or vertical posts must be erected so that they cannot tilt, and must have firm bearing. Inclined shores must be braced securely against slipping or sliding. The bearing ends of shores should be cut square and have a tight fit at splices. Splices must be secure against bending and buckling as provided in Section 1.3.3. Connections of shore heads to other framing should be adequate to prevent the shores from falling out when reversed bending causes upward deflection of the forms.

2.5.2—*Centering*—Centering is the highly specialized falsework used in the construction of arches, shells, space structures, or any continuous structure where the entire falsework is lowered (struck or decentered) as a unit to avoid introducing injurious stress in any part of the structure. The lowering of the centering is generally accomplished by the use of sand boxes, jacks, or wedges beneath the supporting members. For the special problems associated with the construction of centering for folded plates, thin shells, and long span roof structures, see Section 4.5.

2.5.3—*Shoring for composite action between previously erected steel or concrete framing and cast-in-place concrete* (see Section 4.4).

2.6—Adjustment of formwork

2.6.1—*Before concreting*

(a) Tell-tale devices should be installed on supported forms and elsewhere as required to facilitate detection and measurement of formwork movements during concreting.

(b) Wedges used for final alignment before concrete placement should be secured in position after the final check.

---
*Applies to concrete dimensions only, not to positioning of vertical reinforcing bars or dowels.

(c) Formwork must be so anchored to the ores below that upward or lateral movement any part of the formwork system will be evented during concrete placement.

(d) To insure that lines and grades of finhed concrete work will be within the reuired tolerances, the forms must be contructed to the elevation shown on the formvork drawings. If camber is required in the iardened concrete to resist deflection, it should oe so stipulated on the structural drawings. Additional elevation or camber should be provided to allow for closure of form joints, settlements of mudsills, shrinkage of lumber, dead load deflections and elastic shortening of form members.

Where camber requirements may become cumulative, such as in cases where beams frame into other beams or girders at right angles, and at midspan of the latter, the engineer-architect should specify exactly the manner in which this condition is to be handled.

(e) Positive means of adjustment (wedges or jacks) should be provided to permit realignment or readjustment of shores if excessive settlement occurs.

(f) Runways for moving equipment should be provided with struts or legs as required and should be supported directly on the formwork or structural member. They should not bear on or be supported by the reinforcing steel unless special bar supports are provided. The formwork must be suitable for the support of such runways without significant deflections, vibrations or lateral movements.

2.6.2—*During and after concreting*—During and after concreting, but before initial set of the concrete, the elevations, camber, and plumbness of formwork systems should be checked, using telltale devices. *Appropriate adjustments should be promptly made where necessary.* If, during construction, any weakness develops and the falsework shows any undue settlement or distortion, the work should be stopped, the affected construction removed if permanently damaged, and the falsework strengthened.

Formwork must be continuously watched so that any corrective measures found necessary may be promptly taken. Form watchers must always work under safe conditions and should establish in advance a method of communication with placing crews in case of emergency.

2.7—Removal of forms and supports

2.7.1—*Discussion*—Although the contractor is responsible for design, construction, and safety of formwork, it is recommended that time of removal of forms or shores be specified by the engineer-architect because of the possibility of injury to concrete which may not have attained full strength or may be overloaded.

2.7.2—*Recommendations*

2.7.2.1—Suitable tests of job-cured specimens or of concrete in place, methods of evaluating such test results, and minimum standards of strength required should be completely prescribed. The engineer-architect also should specify who will make the cylinders and who will make the tests.

Results of such tests, as well as records of weather conditions and other pertinent information, should be recorded and used by the engineer-architect in deciding when to remove forms.

Determination of the time of form removal should be based on the resulting effect on the concrete.* When forms are stripped there must be no excessive deflection or distortion and no evidence of damage to the concrete, due either to removal of support or to the stripping operation. Where stripping time is less than specified curing time, measures should be taken to provide adequate curing and thermal protection of the stripped concrete. Supporting forms and shores must not be removed from beams, floors, and walls until these structural units are strong enough to carry their own weight and any approved superimposed load, which at no time should exceed the live load for which the floor was designed unless provision has been made by the engineer-architect to allow for anticipated temporary construction loads such as in multistory work (see Section 2.8). In general, removal of forms and supports for suspended structures can be safely accomplished when the ratio of cylinder test compressive strength to design strength is equal to or greater than the ratio of total dead load and construction loads to total design load with a minimum of 50 percent of design compressive strength being required. As a general rule, the forms for columns and piers may be removed before those for beams and slabs. Forms and scaffolding should be designed so they can be easily and safely removed without impact or shock. When quick re-use of forms is desired, formwork should be designed so that it can be removed without removal of sufficient original shores for support until such time as beam or cylinder tests indicate it is safe to remove all shores.

2.7.2.2—When field operations are controlled by the engineer-architect's specifications, the removal of forms, supports, and housing, and

---
*Helpful information on strength development of concrete under varying conditions of temperature and with various admixtures may be found in ACI 306 "Recommended Practice for Cold Weather Concreting" and ACI 605 "Recommended Practice for Hot Weather Concreting."

the discontinuance of heating and curing must follow the requirements of the contract documents. When test beams or cylinders are used to determine stripping times they should be cured under conditions which are not more favorable than the most unfavorable conditions for the portions of the concrete which the test specimens represent. The curing record (time, method, temperature) of the concrete and the weather conditions during the curing of concrete, as well as the test cylinder records, will serve as the basis on which the engineer-architect will determine his approval of form stripping.

2.7.2.3—When field operations are not controlled by the specifications, under ordinary conditions form and supports should remain in place for not less than the following periods of time. These periods represent cumulative number of days or fractions thereof, not necessarily consecutive, during which the temperature of the air surrounding the concrete is above 50 F. If high-early-strength concrete is used, these periods may be reduced as approved by the engineer-architect. Conversely, if low temperature concrete or retarding agents are used, then these periods may be increased at the discretion of the engineer-architect.

Walls*			12-24 hr
Columns*			12-24 hr
Sides of beams and girders*			12-24 hr
Pan joist forms†			
30 in. wide or less			3 days
Over 30 in. wide			4 days
	Where design live load is:		
	<DL	>DL	
Arch centers	14 days	7 days	
Joist, beam, or girder soffits‡			
Under 10 ft clear span between supports	7 days§	4 days	
10 to 20 ft clear span between supports	14 days§	7 days	
Over 20 ft clear span between supports	21 days§	14 days	
Floor slabs‡			
Under 10 ft clear span between supports	4 days§	3 days	
10 to 20 ft clear span between supports	7 days§	4 days	
Over 20 ft clear span between supports	10 days§	7 days	
Post-tensioned slab system**			As soon as full post-tensioning has been applied
Supported slab systems**			Removal times are contingent on reshores, where required, being placed as soon

as practicable after stripping operations are complete but not later than the end of the working day in which stripping occurs. Where reshores are required to implement early stripping while minimizing sag or creep (rather than for distribution of superimposed construction loads as covered in Section 2.8), capacity and spacing of such reshores should be specified by the engineer-architect.

2.8—Shoring and reshoring for multistory structures

2.8.1—*Discussion*—Multistory work presents special conditions particularly in relation to removal of forms and shores. Re-use of form material and shores is an obvious economy. Furthermore, the speed of construction customary in this type of work provides the additional advantage of permitting other trades to follow concreting operations from floor to floor as closely as possible. However, the shoring which supports green concrete is necessarily supported by lower floors which may not be designed for these loads. For this reason shoring must be provided for a sufficient number of floors to develop the necessary capacity to support the imposed loads without excessive stress or deflection.

For purposes of this discussion the following definitions apply:

Shores—Vertical support members designed to carry the weight of formwork, concrete, and construction loads above.

Reshores — Shores placed firmly under a stripped concrete slab or structural member where the original formwork has been removed thus requiring the new slab or structural member to support its own weight and construction loads posted to it. Such reshores are provided to transfer additional construction loads to other slabs or members and/or to impede deflection due to creep which might otherwise occur.

Permanent shores—The original shores supporting forms which are designed for removal without disturbing the original shores or permitting the new concrete to support its own weight and additional construction loads above. Two basic systems are used:

(1) *The king stringer system.* This employs ledgers on the sides of the stringer which may be released permitting the removal of the joist and form contact surfaces between the stringers.

*Where such forms also support formwork for slab or beam soffits, the removal times of the latter should govern.
†Of the type which can be removed without disturbing forming or shoring.
‡Distances between supports refer to structural supports and not to temporary formwork or shores.
§Where forms may be removed without disturbing shores, use half of values shown but not less than 3 days.
**See Section 2.8 for special conditions affecting number of floors to remain shored or reshored.

(2) *The king shore system.* In this system, the stringer is attached to the side of the shores so that the stringer may be removed, permitting the release of the joists and sections of the form contact surfaces. The shores and a trapped strip of contact surface are all that remain in place.

It should be kept in mind that in multistory work with permanent shores, once the lower floor of shores has been removed, the slabs are all in elastic support and therefore capable of deflection in direct proportion to their individual share of loads above as well as to their developed moduli of elasticity. For this reason, if reshores are properly installed, they serve the same function and act in the same capacity as permanent shores.

2.8.1.1.1—*Advantages of the two systems*

Reshores—Stripping formwork is more economically accomplished if all the material can be removed at the same time and moved from the area before placing reshores. No material is trapped which might cause unsightly offsets or joint lines in the exposed soffits. Reshores are installed to conform to the natural distribution of loading on the slab due to its own weight as well as from shores posted to the slab from above.

Permanent shores—Stripping of forms may be accomplished at an earlier age because large areas of concrete are not required to carry their own weight. The cost of the second placing of shores is eliminated. Use of permanent shores avoids the special attention required to assure that shores are placed uniformly tight under the slab. It also provides better assurance that shores are placed in the same pattern on each floor.

2.8.2—*Design* — Recommendations of Section 1.3.3 through 1.3.6 should be followed for reshores or permanent shores.

Shores in the lower stories should be designed to carry the full weight of the concrete and formwork posted to them prior to the removal of the first story of shores supported by the ground. Shores above the first level should be designed to carry a minimum of one and one-half times the weight of a given floor of concrete, forms, and construction loads.*

Where selective reduction in the number of reshores required for lower floors is made, the size of the shore should be carefully determined so as to assure its adequacy for the loads posted thereon.

In determining the number of floors to be shored to support construction loads above, the following factors should be considered:

1. Design load capacity of the slab or member including live load, partition loads, and other loads for which the engineer designed the slab. Where the engineer-architect included allowances for construction loads, such values should be shown on the structural drawings.
2. Dead weight of the concrete and formwork
3. Construction live loads involved, such as placing crews or equipment
4. Design strength of concrete specified
5. Cycle time between placement of successive floors
6. Developed strength of the concrete at the time it is required to support new loads above
7. Span of slab or structural member between structural supports
8. Type of forming systems; i.e., span of horizontal forming components, individual shore loads.

2.8.3—*Placing and removing reshores*

2.8.3.1—Reshoring is one of the most critical operations in formwork; consequently reshoring procedure should be planned in advance and approved by the engineer. Operations should be performed so that at no time will large areas of new construction be required to support combined dead and construction loads in excess of their capability as determined by design load and developed concrete strength at the time of stripping and reshoring. *While reshoring is under way, no construction loads should be permitted on the new construction.*

In no case should reshores be so located as to significantly alter the pattern of stress determined in the structural analysis or to induce tensile stresses where reinforcing bars are not provided.

When placing reshores, care should be taken not to preload the lower floor and also not to remove the normal deflection of the slab above. The reshore is simply a strut and should be tightened only to the extent that no significant shortening will take place under load.

Size and number of reshores must provide a supporting system capable of carrying any loads that may possibly be imposed on it.

Adequate provisions should be made for lateral bracing during this operation. Reshores should be located in the same position on each floor so that they will be continuous in their support from floor to floor. Where the number of reshores on a floor is reduced, such reshores shall be placed directly under a shore position on the floor above. When shores above are not directly over reshores, an analysis should be made to determine whether or not detrimental

---

*See Grundy, Paul, and Kabaila, A., "Construction Loads on Slabs with Shored Formwork in Multistory Buildings," ACI JOURNAL, *Proceedings* V. 60, No. 12, Dec. 1963, pp. 1729-1738.

bending stresses are produced in the slab. Where slabs are designed for light live loads, or on long spans, where the loads on the reshores are heavy, care should be used in placing these reshores, so that the loads on the reshores do not cause excessive punching shear or bending stress in the slab.

When stripping forms before slabs are strong enough to carry their own dead load and/or construction loads above, the following procedures should be followed during stripping and reshoring:

2.8.3.1.1—Reshoring beam and girder construction—The forms should be removed from one girder at a time, and the girder should be reshored before any other supports are removed. After the supporting girders are reshored, the forms should be removed from one beam with its adjacent slabs (unless the forms for slab have already been removed; see Section 2.7.2.3) and the beam should be reshored before any other supports are removed. Each long span slab (clear span 10 ft or more) should be reshored along the center line of the span unless the framing system itself is composed of a clear span between supports. If a line of reshores at midspan is not in line with shores on the floor above, the slab should be checked for its capacity to resist reversal of stresses or punching shear.

Slabs should not be reshored until supporting beams and girders have been reshored.

2.8.3.1.2—Reshoring flat slabs—Shore removal and reshoring should be planned and located to avoid reversal of stresses or inducence of tension in slabs where reinforcement is not provided for design loads. Reshores should be placed along the intersection line of the column strip and the middle strip in both directions. Such reshoring should be completed for each panel as it is stripped before removing forms for adjacent panels. For flat slabs whose column spacing exceeds 25 ft, it is desirable to plan the form construction so that shores at intersections of column strips with middle strips may remain in place during stripping operations.

2.8.3.2 — Removal of reshoring — Reshoring should not be removed until the slab or member supported has attained sufficient strength to support all loads posted to it. Removal operations should be carried out in accordance with a planned sequence so that the structure supported is not subjected to impact or loading eccentricities.

In no case should reshores be removed within 2 days or 2 floors of a freshly placed slab.

CHAPTER 3—MATERIALS FOR FORMWORK

3.1—Discussion

The selection of materials suitable for formwork should be based on maximum economy to the contractor, consistent with safety and the quality required in the finished work. Approval by the engineer-architect, if required, should be based only on safety and quality of finished work.

3.2—Properties of materials

ACI Special Publication No. 4, *Formwork for Concrete*, describes the formwork materials commonly used in the United States and provides extensive related data for form design. Table 3.2 indicates other available sources of design and specification data for formwork materials. This tabulated information should not be interpreted to exclude the use of any other materials which can meet quality and safety requirements established for the finished work.

3.3—Accessories

3.3.1—*Discussion*

3.3.1.1—Form ties—A form tie is a tensile unit adapted to holding concrete forms against the active pressure of freshly placed plastic concrete. In general, it consists of an inside tensile member and an external holding device, both made to specifications of various manufacturers. These manufacturers also publish recommended working loads on the ties for use in form design. There are two basic types of tie rods, the prefabricated rod or band type, and the threaded internal disconnecting type. Their suggested working loads range from 1000 to over 50,000 lb.

3.3.1.2—Form anchors—Form anchors are devices used in the securing of formwork to previously placed concrete of adequate strength. The devices normally are embedded in the concrete during placement. Actual load carrying capacity of the anchors depends on the strength of concrete in which they are embedded, the area of contact between concrete and anchor, and the depth of embedment. Manufacturers publish design data and test information to assist in the selection of proper form anchor devices.

3.3.1.3—Form hangers—Form hangers often are used to support formwork loads from a structural steel or precast concrete framework.

TABLE 3.2—FORM MATERIALS AND STRENGTH DATA FOR DESIGN*

Item	Principle use	Specification and design data sources
Lumber	Form framing, sheathing, and shoring	"National Design Specification for Stress Grade Lumber and Its Fastenings," National Lumber Manufacturers Association ACI Special Publication No. 4, *Formwork for Concrete* *Design data:* *Wood Handbook*, Forest Products Laboratory, U.S. Department of Agriculture *Wood Structural Design Data*, National Lumber Manufacturers Association *Timber Design and Construction Handbook* (in Canada) Part 4, Section 4.3, "Wood," of the National Research Council National Building Code of Canada
Plywood	Form sheathing and panels	U.S. Product Standard PS 1-66 ACI Special Publication No. 4, *Formwork for Concrete* *Design data:* ASTM, USASI CSA Plywood Manufacturers Association of British Columbia American Plywood Association *Plywood: Properties, Design and Construction*
Steel	Heavy forms and falsework	*Manual of Steel Construction*, American Institute of Steel Construction *Light Gage Cold-Formed Steel Design Manual*, American Iron and Steel Institute Manufacturer's specifications and recommendations
	Column and joist forms	Simplified Practice Recommendations R87-32 and R 265-63, U.S. Department of Commerce Manufacturers' specifications and recommendations *Design data: Manual of Standard Practice*, Concrete Reinforcing Steel Institute
	Permanent forms	American Iron and Steel Institute and individual manufacturer's recommendations
	Welding of permanent forms	*Specification:* American Welding Society *Design data:* American Iron and Steel Institute
Aluminum†	Lightweight panels and framing; bracing and horizontal shoring	Manufacturers' handbooks *Aluminum Construction Manual*, the Aluminum Association
Hardboard‡	Form liner and sheathing; pan forms for joist construction	Manufacturers' data CSA

Insulating board:
  Wood fiber
  Glass fiber
  Foamed plastic
	Permanent forms	Manufacturers' data
		CSA

Fiber or laminated paper pressed tubes or forms	Column and beam forms; void forms for slabs, beams, girders, and precast piles	Manufacturers' specifications and recommendations
Corrugated cardboard	Internal and under-slab voids; voids in beams and girders (normally used with internal "egg crate" stiffeners)	Manufacturers' specifications and recommendations
Concrete	Footings	ACI Building Code, ACI 318-63
	Permanent forms	ACI Building Code, ACI 318-63
	Precast floor and roof units	ACI 711-58
	Molds for precast units	—
Fiber-glass-reinforced plastic	Ready-made column and dome pan forms; custom-made forms for special architectural effects	ACI Special Publication No. 4, *Formwork for Concrete* and manufacturers' specifications

*Since handbooks, standards, and specifications of the type cited here are frequently rewritten or updated, the latest available version should be consulted.
†Shall be readily weldable, nonreactive to concrete or concrete containing calcium chloride, and protected against galvanic action at points of contact with steel.
‡Check surface reaction with wet concrete.

(continued on next page)

TABLE 3.2 (cont.)—FORM MATERIALS AND STRENGTH DATA FOR DESIGN

Item	Principal use	Specifications and design data sources

Plastics:
  Polystyrene
  Polyethylene
  Polyvinyl chloride
	Form liners for decorative concrete	Manufacturers' data

Rubber	Form lining and void forms	Manufacturers' data
Form ties, anchors, and hangers	For securing formwork against placing loads and pressures	Manufacturers' specifications; see Section 1.3 of this standard for recommended safety factors ASTM
Plaster	Waste molds for architectural concrete	Manufacturers' recommendations
Coatings	Facilitate form removal	Manufacturers' specifications
Steel joists	Formwork support	"Standard Specifications and Load Tables for Open Web Steel Joists," Steel Joist Institute
Steel frame shoring	Formwork support	"Recommended Steel Frame Shoring Erection Procedure," Steel Scaffolding and Shoring Institute; also manufacturers' data
Form insulation	Cold weather protection of concrete	ACI 306-66 and manufacturers' specifications

3.3.2—*Recommendations*

3.3.2.1—Recommended factors of safety for ties, anchors, and hangers are given in Section 1.3.1. Yield point of the material should not be exceeded.

3.3.2.2—The rod or band type form tie, with supplemental provision for spreading the forms and a holding device engaging the exterior of the form, is the common type used for light construction.

The threaded internal disconnecting type is more often used for formwork on heavy construction such as heavy foundations, bridges, power houses, locks, dams, and architectural concrete.

Removable portions should be of a type which can be readily removed without damage to the concrete and which leave the smallest practicable holes to be filled.

Where ties are permitted in construction of water-retaining structures, they should be designed to prevent seepage or flow of water along the embedded tie.

Although there is no general uniformity at the present time, a minimum specification for form ties should require that the bearing area of external holding devices be adequate to prevent severe crushing of form lumber.

3.3.2.3—Form hangers must support the dead load of forms, weight of concrete, and construction and impact loads. Form hangers should be symmetrically arranged on the supporting member to minimize twisting or rotation of supporting members.

3.3.2.4—Where the concrete surface is to be exposed and appearance is important, the proper type of form tie or hanger which will not leave exposed metal at the surface is essential.

3.4—Form coatings or release agents

3.4.1—*Form coating*—Form coatings or sealers are usually applied in liquid form to contact surfaces either during manufacture or in the field to serve one or more of the following purposes:
  (a) Alter the texture of the contact surface
  (b) Improve the durability of the contact surface
  (c) In addition to (b) above, to facilitate release from concrete during stripping
  (d) Seal the contact surface from intrusion of moisture.

3.4.2—*Release agents*—Form release agents are applied to the form contact surfaces to prevent bond and thus facilitate stripping. They may be applied permanently to form materials in manufacture or applied to the form before each use. When applied in the field before each use, care must be exercised to prevent coating adjacent construction joint surfaces or reinforcing steel.

3.4.3—*Manufacturers' recommendations*—Manufacturers' recommendations should be followed in the use of coatings, sealers, and release agents, but independent investigation of their performance is recommended before use. Where surface treatments such as paint, tile adhesive, or other coatings are to be applied to formed concrete surfaces, be sure that adhesion of such surface treatments will not be impaired or prevented by use of the coating, sealer, or release agent.

REQUIREMENTS FOR SHORING FORMWORK
APPENDIX B

Shoring of concrete formwork is extensive on many concrete construction sites. Proper erection and maintenance of shoring is essential for form and worker safety. To promote safety in shoring concrete formwork, the Scaffolding and Shoring Institute has developed *Recommended Safety Requirements for Shoring Concrete Formwork*. It is reprinted here to provide an authoritative reference.

FOREWORD

The Recommended Safety Requirements for Shoring Concrete Formwork was developed to fill the need for a single, authoritative source of information on shoring as used in the construction industry, and to provide recommendations for the safe and proper use of this type of equipment.

The following recommended safety requirements were designed to promote safety and the safe use of shoring in the construction industry. This document does not purport to be all inclusive or to supplant or replace other additional safety and precautionary measures.

It is suggested that you refer to the Single Post Shore Safety Rules, Steel Frame Shoring Safety Rules, Horizontal Shoring Beam Safety Rules and the Recommended Steel Frame Shoring Erection Procedure. All of these publications are available from the Scaffolding & Shoring Institute.

This document was prepared by the Engineering Committee of the Scaffolding & Shoring Institute, 2130 Keith Building, Cleveland, Ohio 44115.

TABLE OF CONTENTS

INTRODUCTION — Page 269

Section 1 - Definitions — Page 269

Section 2 - General Requirements for Shoring — Page 272

Section 3 - Tubular Welded Frame Shoring — Page 275

Section 4 - Tube and Coupler Shoring — Page 276

Section 5 - Single Post Shores — Page 277

Section 6 - Horizontal Shoring Beams — Page 280

RECOMMENDED SAFETY REQUIREMENTS FOR SHORING CONCRETE FORMWORK

INTRODUCTION

Because of the widespread and constantly growing use of shoring in construction today, it is vital that it be used properly and safely.

At present, the subject of shoring is not covered in most standard reference works on safe construction practices, therefore, two objectives were established before preparing these Recommended Safety Requirements.

1. To fill the need for a single, authoritative source of information on shoring and the safe and proper use thereof, and -

2. To provide a guide to various Federal, State, and local authorities having jurisdiction over construction work in developing their own codes.

However, the information and recommendations contained herein do not supersede any applicable Federal, State, or local code, ordinance, or regulation.

SECTION 1

DEFINITIONS

1.1 ACCESSORIES - Those items other than frames, braces, horizontal shoring beams fixed or adjustable, or post shores used to facilitate the construction of shoring.

1.2 ADJUSTABLE BEAMS - Beams whose length can be varied within predetermined limits.

1.3 ADJUSTMENT SCREW - A leveling device or jack composed of a threaded screw and an adjusting handle used for the vertical adjustment of the shoring and formwork.

1.4 ALLOWABLE LOAD - The ultimate load divided by factor of safety.

1.5 BRACKET - A device used on scaffold to extend the width of the working platform.

1.6 BASE PLATE - A device used to distribute the leg load.

REQUIREMENTS FOR SHORING FORMWORK

1.7 CLEAR SPAN - The distance between facing edges of members used to support horizontal shores.

1.8 COUPLER OR CLAMP* - A device used for fastening together tubular component parts.

1.9 COUPLING PIN - An insert device used to align lifts or tiers vertically.

1.10 CROSS BRACING - A system of members which connect frames or panels of shoring laterally to make a tower or continuous structure.

1.11 DEAD LOAD - The load of forms, stringers, joists, reinforcing rods, and the actual concrete to be placed.

1.12 DESIGN STRESS - Allowable stress for stress-grade lumber conforming to the recommended unit stress indicated in the tables in the "Wood Structure Design Data Book," by National Forest Products Association (Formerly National Lumber Manufacturers Association), Washington, D.C. For form lumber design, see Paragraph 2.8.

1.13 EXTENSION DEVICE - Any device, other than an adjustment screw, used to obtain vertical adjustment of shoring towers.

1.14 FACTOR OF SAFETY - The ratio of ultimate load to the allowable load.

1.15 FORMWORK - The material used to give the required shape and support of poured concrete, consisting primarily of:

> Sheathing - material which is in direct contact with the concrete such as wood, plywood, metal sheet or plastic sheet.
>
> Joists - members which directly support sheathing.
>
> Stringers or ledgers - members which directly support the joists, usually wood or metal load-bearing members.

1.16 FRAME OR PANEL* - The principal prefabricated, welded strucural unit in a tower.

1.17 HORIZONTAL SHORING BEAM - Adjustable or fixed length span members, either beam or truss type used as joists, stringers or ledgers.

1.18 JOISTS - See Formwork.

*(These terms and others so marked may be used synonymously.)

REQUIREMENTS FOR SHORING FORMWORK

1.19 **LIFTS OR TIERS*** - The number of frames or shores erected one above each other in a vertical direction.

1.20 **LIVE LOAD** - The total weight of workmen, equipment, buggies, vibrators, and other loads that will exist and move about due to the method of placement, leveling and screeding of the concrete pour.

1.21 **LOCKING DEVICE**

    a. A device used to secure the cross brace to the frame or panel.

    b. A device used to lock an adjustable beam at a specific (fixed length).

1.22 **NON-ADJUSTABLE BEAMS** - Beams of fixed length.

1.23 **POST SHORE OR POLE SHORE*** - Individual vertical member used to support loads.

    a. **Adjustable Timber Single Post Shore** - Individual wooden timbers used with fabricated clamps to obtain adjustment.

    b. **Fabricated Single Post Shores**

        Type I - Single all-metal post, with a fine adjustment screw or device in combination with pin and hole adjustment or clamp.

        Type II - Single or double wooden post upper members with metal clamp or screw and usually manufactured as a complete unit.

    c. **Timber Single Post Shores** - Wood timber used as a strutural member for shoring support.

1.24 **RESHORING** - A system used during the construction operation in which the original shores are removed and replaced in a sequence planned to avoid any damage to partially cured concrete, and to aid in the support of additional construction.

1.25 **SAFE LEG LOAD** - The load which can safely be directly imposed on the frame leg. (See Allowable Load).

1.26 **SHOCK LOAD** - Impact of material such as the concrete as it is released or dumped during placement.

1.27 **SHORE HEADS** - Flat or formed metal pieces which are placed and centered on vertical shoring member.

1.28 **SHORING LAYOUT** - A properly designed drawing prepared prior to erection showing arrangement of equipment for proper shoring.

1.29 SILL OR MUD SILL* - A footing (usually wood) which distributes the vertical shoring loads to the ground or slab below.

1.30 SPACING - The horizontal distance between support members.

1.31 STRINGERS OR LEDGERS* - See Formwork.

1.32 TOWERS - A composite structure of frames, braces, and accessories.

1.33 TUBE AND COUPLER SHORING - An assembly used as a load-carrying structure consisting of tubing or pipe which serves as posts, braces, and ties, a base supporting the posts, and special couplers which serve to connect the uprights and join the various members.

1.34 ULTIMATE LOAD - The load at which the structure fails.

SECTION 2

GENERAL REQUIREMENTS FOR SHORING

2.1 Shoring installations constructed in accordance with these recommended safety requirements shall have a properly designed shoring layout.

2.2 The shoring layout shall include details, accounting for unusual conditions such as heavy beams, sloping areas, ramps and cantilevered slabs, as well as plan and elevation views.

2.3 A copy of the shoring layout shall be available and used on the job site at all times.

2.4 The shoring layout shall be prepared or approved by a person qualified to analyze the loadings and stresses which are induced during the construction process.

2.5 The minimum total design load for any formwork and shoring used in a slab and beam structures shall be not less than 100 lbs. per square foot for the combined live and dead load regardless of slab thickness; however, the minimum allowance for live load shall be not less than 20 lbs. per square foot.

2.6 When motorized carts or buggies are used, the design load, as described in Section 2.5, shall be increased 25 lbs. per square foot.

2.7 Allowable loads shall be based on a safety factor consistent with the type of shoring used and as set forth in Sections 3, 4, 5 and 6.

2.8 The design stresses for form lumber and timbers shall be commensurate with the grade, conditions, and species of lumber used, in accordance with the latest edition of the National Design Specification for Stress-Grade Lumber and its Fastenings (National Forest Products Association).

2.9 Design stresses may be increased for short term loading conditions as provided in "The Wood Structure Design Data Book" (National Forest Products Laboratory).

2.10 The design stresses used for form lumber and timber shall be shown on all drawings, specifications and shoring layouts.

2.11 The sills for shoring shall be sound, rigid and capable of carrying the maximum intended load without settlement or displacement. The load should be applied to the sill in a manner which will avoid overturning of the tower, or the sill.

2.12 When shoring from soil, an engineer or other qualified person shall determine that the soil is adequate to support the loads which are to be placed on it.

2.13 Precautions should be taken so that weather conditions do not reduce the load-carrying capacity of the soil below the design minimum.

2.14 When shoring from fill or when excessive earth disturbance has occurred, an engineer or other qualified person shall supervise the compaction and reworking of the disturbed area, and determine that it is capable of carrying the loads which are to be imposed on it.

2.15 Suitable sills shall be used on a pan or grid dome floor or any other floor system involving voids, where vertical shoring equipment could concentrate an excessive load on a thin concrete section.

2.16 Formwork, together with shoring equipment, shall be adequately designed, erected, braced and maintained so that it will safely support all vertical and lateral loads that might be applied, until such loads can be supported by the concrete structure.

2.17 Construction requirements for forming shall be in accordance with the provisions of the current issue of "Recommended Practice for Concrete Formwork" published by the American Concrete Institute.

2.18 When temporary storage of reinforcing rods, material, or equipment, on top of formwork becomes necessary, special consideration shall be given to these areas and they shall be strengthened to meet these loads.

REQUIREMENTS FOR SHORING FORMWORK

2.19 All shoring equipment shall be inspected by the contractor who erects the equipment prior to erection to determine that it is as specified in the shoring layout.

2.20 Damaged equipment shall not be used for shoring.

2.21 Erected shoring equipment shall be inspected by the contractor who is responsible for placement of concrete immediately prior to pour, during and after pour, until concrete is set.

2.22 If any deviation is necessary because of field conditions, the person who prepared the shoring layout shall be consulted for his approval of the actual field setup before concrete is placed, and the shoring layout shall be revised to indicate any approved changes.

2.23 The shoring setup shall be checked by the contractor who erects the equipment to determine that all details of the layout have been met.

2.24 The completed shoring setup shall be a homogeneous unit or units and shall have the specified bracing to give it lateral stability.

2.25 The shoring setup shall be checked by the contractor who erects the equipment or make sure that bracing specified in the shoring layout for lateral stability is in place.

2.26 All vertical shoring equipment shall be plumb in both directions, unless otherwise specified in the layout, the maximum allowable deviation from the vertical is 1/8 inch in 3 feet, but shall not exceed 1 inch in 40 feet. If this tolerance is exceeded, the shoring equipment shall not be used until readjusted within this limit.

2.27 Any erected shoring equipment, upon inspection, that is damaged or weakened, shall be immediately removed and replaced by adequate shoring.

2.28 Loaded shoring equipment shall not be released or removed until the approval of a qualified engineer has been received.

2.29 Release and removal of loads from shoring equipment shall be planned so that the equipment which is still in place is not overloaded.

2.30 Slabs or beams which are to be reshored should be allowed to take their actual permanent deflection before final adjustment of reshoring equipment is made.

2.31 While the reshoring is underway, no construction loads shall be permitted on the partially cured concrete.

REQUIREMENTS FOR SHORING FORMWORK   275

2.32 The allowable load on the supporting slab shall not be exceeded when reshoring.

2.33 The reshoring shall be thoroughly checked by the architect/ engineer to determine that it is properly placed and that it has the allowable load capacity to support the areas that are being reshored.

SECTION 3

TUBULAR WELDED FRAME SHORING

3.0 See Section 2, "General Requirements for Shoring"

3.1 Metal tubular frames used for shoring shall have allowable loads based on tests conducted according to a standard test procedure for the Compression Testing of Scaffolds and Shores as established by the Scaffolding & Shoring Institute or its equivalent.

3.2 Design of shoring layouts using tubular welded frames shall be based on allowable loads which were obtained using these test procedures, and at least a 2.5 to 1 safety factor.

3.3 All metal frame shoring equipment shall be inspected by the contractor who erects the equipment before erection.

3.4 Metal frame shoring equipment and accessories shall not be used if excessively rusted, bent, dented, rewelded beyond the original factory weld locations, having broken welds, or having other defects.

3.5 All components shall be in good working order and in a condition similar to that of original manufacture.

3.6 When checking the erected shoring frames with the shoring layouts, the spacing between towers and cross brace spacing shall not exceed that shown on the layout, and all locking devices shall be in the closed position.

3.7 Devices to which the external lateral stability bracing are attached shall be securely fastened to the legs of the shoring frames.

3.8 Base plates, shore heads, extension devices, or adjustment screws shall be used in top and bottom of each leg of every shoring tower.

3.9 All base plates, shore heads, extension devices or adjustment screws shall be in firm contact with the footing sill, and/or the form material, and shall be snug against the legs of the frame.

3.10 There shall be no gaps between the lower end of one frame and the upper end of the other frame.

3.11 Any component which cannot be brought into proper alignment or contact with the component into or onto which it is intended to fit shall be removed and replaced.

3.12 When two or more tiers of frames are used, they shall be cross braced. Towers shall have lateral bracing in accordance with manufacturers recommendations and as shown on the shoring layout.

3.13 Eccentric loads on shore heads and similar members shall be avoided.

3.14 Special precautions shall be taken when formwork is at angles or sloping or when the surface shored from is sloping.

3.15 Adjustment screws shall not be adjusted to raise formwork after the concrete is in place.

3.16 Brackets (Section 1.5) shall not be used for carrying shoring loads.

SECTION 4

TUBE AND COUPLER SHORING

4.0 See Section 2, "General Requirements for Shoring"

4.1 Tube and coupler towers used for shoring shall have allowable loads based on tests conducted according to a standard test procedure for the Compression Testing of Scaffolds and Shores such as the one established by the Scaffolding & Shoring Institute, or its equivalent.

4.2 Design of tube and coupler towers in shoring layouts shall be based on allowable loads which were obtained using these test procedures, and at least a 2.5 to 1 safety factor.

4.3 All tube and coupler components shall be inspected by the contractor who erects the equipment before being used.

4.4 Tubes of shoring structures shall not be used if excessively rusted, bent, dented, or having other defects.

4.5 Couplers (clamps) shall not be used if deformed, broken having defective or missing threads on bolts or other defects.

4.6 The material used for the couplers (clamps) shall be of a structural type such as drop-forged steel, malleable iron, or structural grade aluminum. Gray cast iron shall not be used.

4.7 When checking the erected shoring towers with the shoring layout, the spacing between posts shall not exceed that shown on the layout, and all interlocking of tubular members and tightness of couplers should be checked.

4.8 Base plates, shore heads, extension devices, or adjustment screws shall be used in top and bottom of each leg of every shoring tower.

4.9 All base plates, shore heads, extension devices, or adjustment screws shall be in firm contact with the footing sill and/or the form material, and shall be snug against the posts.

4.10 Any component which cannot be brought into proper alignment or contact with the component into or onto which it is intended to fit shall be removed and replaced.

4.11 Eccentric loads on shore heads and similar members shall be avoided.

4.12 Special precautions shall be taken when formwork is at angles or sloping or when the surface shored from is sloping.

4.13 Adjustment screws shall not be used to raise formwork after the concrete is in place.

SECTION 5

SINGLE POST SHORES

5.0 See Section 2, "General Requirements for Shoring"

5.1 When checking erected single post shores with the shoring layout, the spacing between shores in either direction shall not exceed that shown on the layout, and all clamps, screws, pins, and all other components should be in the closed or engaged position.

5.2 For stability, single post shores shall have adequate bracing provided in both the longitudinal and transverse directions, and adequate diagonal bracing shall be provided.

5.3 Devices to which the external lateral stability bracing are attached shall be securely fastened to the single post shores.

5.4 All base plates or shore heads of single post shores shall be in firm contact with the footing sill and form material.

5.5 Use of single post shores in more than one tier is not recommended. If post shores are tiered, consult manufacturer of posts for erection and bracing requirements, or use a layout designed and inspected by a professionsl engineer.

REQUIREMENTS FOR SHORING FORMWORK 278

5.6   Eccentric loads on shore heads shall be avoided.

5.7   Special precautions shall be taken when formwork is at angles, or sloping, or when the surface shored from is sloping.

5.8   Adjustment of single post shores to raise formwork shall not be made after concrete is in place.

5.9   Fabricated single post shores.

   5.9.1   All fabricated single post shores used for shoring shall have working load ratings based on tests conducted according to a standard test procedure established by the Scaffolding & Shoring Institute for fabricated single post shores or its equivalent.

   5.9.2   Design of fabricated single post shores in shoring layouts shall be based on working loads which were obtained using these test procedures and at least a 3 to 1 safety factor.

   5.9.3   All fabricated single post shores shall be inspected by the contractor who erects the equipment before being used.

   5.9.4   Fabricated single post shores shall not be used if excessively rusted, bent, dented, rewelded beyond the original factory weld locations, have broken welds, if they contain timber, they shall not be used if timber is split, cut, has sections removed, is rotted or otherwise structurally damaged.

   5.9.5   All clamps, screws, pins, threads, and all other components shall be in a condition similar to that of original manufacture.

5.10 Adjustable timber single post shores.

   5.10.1 The adjustable timber single post shores shall have working load ratings based on tests conducted according to a standard test procedure established by the Scaffolding & Shoring Institute for fabricated single post shores or its equivalent.

   5.10.2 Timber used shall have allowable working load for each size grade species and shoring height as recommended by the clamp manufacturer.

   5.10.3 Design of an adjustable timber single post shoring layout shall be based on working loads which were

REQUIREMENTS FOR SHORING FORMWORK

obtained using the test procedure and at least a 3 to 1 safety factor.

5.10.4 All timber and adjusting devices to be used for adjustable timber single post shores shall be inspected by the contractor who erects the equipment before erection.

5.10.5 Timber shall not be used if it is split, cut, has sections removed, is rotted, or is otherwise structurally damaged.

5.10.6 Adjusting devices shall not be used if excessively rusted, bent, dented, rewelded beyond the original factory weld locations, have broken welds.

5.10.7 Hardwood wedges shall be used to obtain final adjustment and firm contact with footing sills and form material.

5.10.8 All nails used to secure bracing on adjustable timber single post shores shall be driven home and bent over if possible.

5.11 Timber single post shores.

5.11.1 Timber used as single post shores shall have the safety factor and allowable working load for each grade and species as recommended in tables for wooden columns in the "Wood Structural Design Data Book," prepared by the National Forest Products Association, (Formerly National Lumber Manufacturers Association), Washington, D.C.

5.11.2 Design of timber single post shoring layout shall be based on working loads which were obtained by using the tables referred to in 5.11.1.

5.11.3 All timber to be used for single post shoring shall be inspected by the contractor who erects the equipment before erection.

5.11.4 Timber shall not be used if it is split, cut, has sections removed, is rotted, or is otherwise structurally damaged.

5.11.5 Hardwood wedges shall be used to obtain final adjustment and firm contact with footing sills and form material.

5.11.6 All nails used to secure bracing on timber single post shores shall be driven home and bent over if possible.

SECTION 6

HORIZONTAL SHORING BEAMS

6.0   See Section 2, "General Requirements for Shoring"

6.1   Horizontal shoring beams shall have allowable loads based on tests conducted according to a standard test procedure for testing of horizontal shoring beams established by the Scaffolding & Shoring Institute, or its equivalent.

6.2   The design of horizontal shoring components in shoring layouts shall be based on allowable loads which were obtained using these test procedures, and at least a 2 to 1 safety factor.

6.3   All horizontal shoring beams shall be inspected by the contractor who erects the equipment before using.

6.4   Erected horizontal shoring beams shall be inspected to be certain that the span, spacing, types of shoring beams, and the types' size, heights, and spacing of vertical shoring supports are in accordance with the shoring layout.

6.5   Adequate support shall be provided and maintained to properly distribute shoring loads. When supporting horizontal shoring beams on:

  6.5.1   Masonry walls, they shall have adequate strenth. Brace walls as necessary.

  6.5.2   Ledgers supported by walls using bolts, or other means, the ledgers shall be properly designed and installed per recommendation of supplier or job architect/engineer.

  6.5.3   Formwork, such formwork shall be designed for the additional loads imposed by the shoring beams.

  6.5.4   Structural Steel Framework, the ability of the steel framework to support this construction loading shall be checked and approved by the responsible project architect/engineer.

  6.5.5   Steel hangers, bearing ends of the horizontal shoring beams shall be fully engaged on the hangers. The hangers shall be designed to conform to the bearing end, and to safely support the shoring loads imposed. (Hanger manufacturerers' recommendations shall be followed).

6.6   Precaution shall be taken in the design and the installation of horizontal shoring beams.

REQUIREMENTS FOR SHORING FORMWORK

- 6.6.1 When sloped or supported by sloping ledgers (stringers).

- 6.6.2 When ledger (stringer) height/width ratio exceed 2 1/2 to 1. Under no circumstances shall horizontal shoring beams bear on a single "two by" ledger (stringer).

- 6.6.3 When eccentric loading conditions exist.

- 6.6.4 When ledger (stringer) consists of multiple members. (i.e., double 2x6, 2x8 etc.)

6.7 Bearing ends of horizontal shoring beams shall be properly supported and locking devices properly engaged before placing any load on beams.

6.8 Horizontal shoring beams shall not be supported other than at the bearing prongs unless recommended by supplier.

6.9 Do not nail beam bearing prongs to ledger.

6.10 Adjustable horizontal shoring beams shall not be used as part of a reshoring system.

6.11 Adjustable horizontal shoring beams shall not be used as a stringer (ledger) for other horizontal shoring beams.

MANUFACTURERS OF MATERIALS AND EQUIPMENT
APPENDIX C

Appendix C is included as an aid in locating additional information on some of the building products and concrete forming systems discussed in the text. Many of the companies and associations listed have representatives in cities across the country and may be found in the yellow pages of the telephone directory. However, if none is listed in the directory, a letter to the home office will elicit the necessary information. The following list is divided into two parts: Building Materials, and Concrete Forming-Equipment and Services.

BUILDING MATERIALS

Alliance Wall Corp.
P.O. Box 247
Alliance, Ohio 44601

American Concrete Institute
P.O. Box 4754, Redford Station
Detroit, Michigan 48219

American Plywood Association
1119 A Street
Tacoma, Washington 98401

Georgia-Pacific
Gypsum Division
P.O. Box 311
Portland, Oregon 97207

Gold Bond Building Products
Division of National Gypsum Co.
325 Delaware
Buffalo, N.Y. 14202

Masonite Corporation
29 Wacker Drive
Chicago, Illinois 60606

Owens-Corning Fiberglas Corp.
Fiberglas Tower
Toledo, Ohio 43601

Southern Forest Products Association
P.O. Box 52468
New Orleans, Louisiana 70150

United States Gypsum
101 South Wacker Drive
Chicago, Illinois 60606

Western Wood Products Association
1500 Yeon Building
Portland, Oregon 97204

CONCRETE FORMING – EQUIPMENT AND SERVICES

AllenForm
Con-Form Equipment Corp.
225 N. Arlington Heights Rd.
Elk Grove, Illinois 60007

Bergan Built, Inc.
Subsidiary of Sonoco Products Co.
Box 747
Centralia, Illinois 62801

Ceco Corporation
5601 W. 26th St.
Chicago, Illinois 60650

David White Instruments
Division of Realist, Inc.
N 93 W 16288 Megal Drive
Menomonee Falls, Wisconsin 53051

The Dayton Sure-Grip & Shore Co.
721 Richard St.
Miamisburg, Ohio 45342

Economy Forms Corp.
Box D, East 14th St. Station
Des Moines, Iowa 50316

Gates and Sons, Inc.
90 South Fox St.
Denver, Colorado 80223

Gateway Building Products
3233 W. Grand Ave.
Chicago, Illinois 60651

Jahn Concrete Forming Equipment
6359 E. Evans St.
Denver, Colorado 80222

Patent Scaffolding Co.
Division of Harsco Corp.
11-11 34th Ave.
Long Island City, N.Y. 11106

Richmond Screw Anchor Co.
500 E. 132nd St.
Bronx, N.Y. 10454

Safway Steel Products, Inc.
6228 W. State St.
Milwaukee, Wisconsin 53213

Sonoco Products Co.
Hartsville, South Carolina 29550

Spanall Co.
Division of Patent Scaffolding Co.
11-11 34th Ave.
Long Island City, N.Y. 11106

Symons Manufacturing Co.
200 E. Touhy Ave.
Des Plaines, Illinois 60018

Universal Form Clamp Co.
1238 Kostner Ave.
Chicago, Illinois 60651

SELECTED BIBLIOGRAPHY

American Concrete Institute. *Recommended Practice for Concrete Formwork.* American Concrete Institute, Detroit, Michigan, 1968.

American Plywood Association. *How to Buy and Specify Plywood.* American Plywood Association, Tacoma, Washington, 1966.

——. *Plywood Construction Systems for Commercial and Industrial Buildings.* American Plywood Association, Tacoma, Washington.

——. *Plywood for Concrete Construction.* American Plywood Association, Tacoma, Washington, 1971.

Badzinski, Stanley, Jr. *Carpentry in Residential Construction.* Prentice-Hall, Inc., Englewood Cliffs, New Jersey, 1972.

——. *Stair Layout.* American Technical Society, Chicago, Illinois, 1971.

Cooper, George H., and Badzinski, Stanley, Jr. *Building Construction Estimating.* 3rd ed., McGraw-Hill Book Co., New York, 1971.

Delmar Publishers. *Concrete Form Construction.* Delmar Publishers, Albany, New York, 1946.

Durbahn, W. E. and Sundberg, E. W. *Fundamentals of Carpentry*, Vol. 2, 4th ed. American Technical Society, Chicago, Illinois, 1969.

Huntington, W. C. *Building Construction*. 3rd ed., John Wiley & Sons, Inc., New York, 1963.

Hurd, M. K. *Formwork for Concrete*, 3rd ed., American Concrete Institute, Detroit, Michigan, 1973.

Peurifoy, R. L. *Formwork for Concrete Structures*. McGraw-Hill Book Co., New York, 1964.

Portland Cement Association. *Design and Control of Concrete Mixes*. Portland Cement Association, Chicago, Illinois, 1970.

Stahl, Allen. *NACA Ceiling Systems Handbook*. National Acoustical Contractors Association, Glenview, Illinois, 1967.

United States Gypsum Co. *Drywall Construction Handbook*. United States Gypsum Co., Chicago, Illinois, 1968.

Watson, Don A. *Construction Materials and Processes*. McGraw-Hill Book Co., New York, 1972.

INDEX

Numerals in italics refer to illustrations on those pages

A

Acoustical ceiling, 209-217
  adhesive application, 214-215, *215*
  exposed grid, 216-217, *217*
  horizontal layout, 211-213, *212*
  preparation, 210-211
  types, 210
  vertical layout, 213-214, *214*
  Z-spline, 215-216, *215*, 216
Adjustable brace, *53*
AllenForm, 64-66, *64*, 65
Arch centers, 186-188, *186*, 187

B

Batterboards, 18-20, *19*, 20
Beam forms, 107-120
  bottoms, 113-114, *114*
  hangers, 117-119, *118*, 119
  reshoring, 120
  sides, 115-116, *115*
    installation, 116
    ledger, 116, *116*
  shore removal, 120
  shoring, 107-113
  spandrel, 117, *117*
  stripping, 119-120
  typical, *108*
Braces, function, 52
Building codes, 1-2
Building lines:
  layout, 28-29, *28*
  staking out, 32-33, *33*
Building materials, product information, 282-283

C

Cabinetry, 242-248
  base, 246-247, *246*, 247
  installation, 230
  wall, 244-245, *244*, 245
  wardrobe, 242-243, *243*
  workbenches, 247-248, *248*
Cabinets, 6-7, *6*
Carpenter work, 1

Coil ties, 42-43, *43*, 44
Column, concrete placement, 88
Column footing forms, 8-14, *8*, 9
   battered, 12-14, *12*, 13
   battered form, anchoring, 14, *13*
   circular, 11, *11*
   fiber, 11
   rectangular and square, 9-10
   steel, 11
   stepped, 14
Column forms, 87-105
   adjustable clamps, 94, *93*, (table) 94
   board, *89*
   bracing, 94-95, 105, *96*, 104
   clamping, 90-92, *97*
   cleaning, 80-82
   fiber, 98-100
      bracing, 98, *98*
      cutting, 99
      details, *99*, 100
      obround, 100
      protection, 98
      size table, 98
      stripping, 98-99
      template, 98, *98*
   installation, 92-95
   panel, 101-102, *101*
   plumbing, 94-95, *95*
   plywood, 89-92, *90*
   pressures, 87-88, (table) 88
   rectangular, 90, *91*
   simple, 96-97, *96*
   spreaders, 105, *104*
   steel, 102-105, *103*
      bracing, *104*
      obround, 102, *102*
   steel ply, 101-102, *101*
   template, 92-94, *93*
   Uni-Form, 102
   wood, 88-97
Commercial construction, definition, 1
Concrete forming, 2-4
   equipment suppliers, 283-285
   ribbed and waffle slab, 4
   systems, 3-4

Concrete formwork standards, 250-266
   accessories, 264-266
   adjustments, 260-261
   centering, 260
   coatings, 266
   design recommendations, 252-256
   falsework, 260
   general practices, 257
   material properties, 264-266
   removal, 260-261
   reshoring, 262-264
   safety precautions, 256-257
   shoring, 262-264
   tolerances, 258-260
   workmenship, 257-258
Cornices, decorative, 230-235, *231*
   fitting inside corners, 233-234, *233*
   fitting outside corners, 234-235, *234*

D

Design tables:
   concrete pressure, 48
   joists and studs, 49
   plywood, 48
   wales, 50
Display case, cornice height, 236-237, *236*, 237
Display racks, 235, *235*
Door bucks, 183-186
Door frames, 183-186
   steel, 184, *184*
   wood, 185-186, *185*

E

Economy form (*see* Flat slab form)

F

Fasteners, 232-233, *232*
Fireproofing steel beams, 117-119, *118*, 119
Fixtures, 6-7, 237-242
   counters, free standing, 239, *239*

display case, cornice height, 240, *240*
display case, counter high, 238-239, *238*, 239
display platform, 241, *242*
installation, 230
shelves, 241, *241*
Flat slab forms, 132-147
  Economy Form, 144-147
    erection, 144-146, *144*, 145
    stripping, 146-147, *146*, 147
  Symons slab shore, 133-144, *134*
    assembly, 134-137, *135*, 136, 137
    erection, 137-141, *139*, 140, 141
    layout, 138, *138*
    preparation, 134-138
    stripping, 142-144, *142*, 143
Floor forms, 122-153
  Ellis shore, 123, *124*, 125
  shoring, 122-132, *123*, 124, 125, 127, 129, 130, 131, 132
  stripping, 133
Floor slabs (*see* Floor forms)
Flying shores, 129-130, *129*
Form aligners, 79-80, *80*, 81
Form fabricating bench, 53
Forms for slabs on grade, 18
Formwork:
  building, 3, *2*
  design, 48-50
  erection, 3
  materials, 35-46
  stripping, 3

G

Gang forms, 82-84, *82*
Girder forms (*see* Beam forms)
Grade elevations, 20-21
  measuring, 26-28, *27*
  transferring, 27-28, *28*
Grounds, plaster, 202-203, *203*
Gypsum wallboard, 203-204, *203*, 204

H

Hardware, concrete form, 40-46

Horizontal shoring (See Shoring, horizontal)

J

Job built forms, 3, 50-66

K

Kicker, 107, *108*

L

Ledger, 107, *108*
Level-transit, 21-32
Locating footings, 18-21
Lumber, 36-37

M

Movable partitions, 5-6 (*see also* Demountable partitions)
Mudsills, 111-113, *112*

N

Nails, 39-40

O

Openings in wall forms, 55-64
  core box, 56, *56*
  door bucks, 62-64, *62*, 63
  window form filler strip, 60-61, *60*, 61
  windows, 57-62, *57*, 59
  tube inserts, *55*

P

Paneling, 203-209
  corner finishing, 207-209, *208*
  installation, 204-207, *205*, 206, 207
  installation procedure, 207-209
  prefinished, 204-209
  scribing, 207, *207*

Partitions, 189-203
  demountable, 219-220, 221-229
    ceiling height, 221-224, *222*, 223, 224
    cornice height, 225-226, *226*
    installing, 227-229
    rail height, 227, *227*, 228
  door openings, 190-192, *191*, 192
  furring, 200-202, *201*, 202
  header height, 194-195, *195*
  header size, 192-194, *193*
  metal stud, 198-200, *198*, 199, 200
  movable, 219-229
    installation, 220-221
    layout, 220
  pipe chase, 200, *200*
  plate layout, 197-200
  story pole, 190, *190*
  window openings, 195-196, *196*, 197
  wood stud, 189-190

P

Perimeter work, 6-7
Plaster grounds, 202-203, *203*
Plate, concrete form, 52
Plate layout, 171, *171*
Plywood, 37-39
Pressures on formwork, 46-50

R

Release agents, 39
Resilient furring, 200-202, *202*
Ribbed slabs:
  flange form installation, 147-148, *148*
    stripping, 149
  form materials, 147
  long form, 149-150, *149*
    installation, 150
    stripping, 150
Rod clamp tie, 45-46, *46*

S

She bolt, 43, *44*, 45

Sheathing, form 51
Shoring:
  bracing, 113
  definitions, 269-272
  double post, 107
  erection, 113
  flying, 129-130, *129*
  general requirements, 272-275
  horizontal, 130-132, 280-281, *130*, 131, 132
  inspection, 126, 128-129
  single post, 107-110, 277-279, *109*, 110
  steel, 278
  timber, 279
  timber, adjustable, 278
  spacing, 113
  steel frame, 126-129, *127*, 129
  tube and coupler, 276-277
  tubular steel, 107, 275-276, *111*
Snap ties, 40-42, *41*, 42
Spandrel beam form, 117, *117*
Spreaders, function, 52
Stair:
  between walls, 163-167
  example, 157-158, *157*
  form, 4, *164*, 165, 167
  form removal, 163
  forming free standing, 168-169
  grade, 158-159, *159*
  layout, 155-158
  layout procedure, 165-166
  riser forms, 162-163, *161*, 162, 166
  section, typical, *156*
  stringer layout, 160-161, *160*
  stringer placement, 161, *160*
Stakes, function of, 52
Steel panel forms, 76-82
  erection, 76-79, *76*, 77, 78, 79
Store fronts, 4-5, *4*
  building, 170-175
  colored panels, 179-181, *180*, 181
  display window, 173
  door frame, 173-174, *174*
  exterior finish, 175-181
  plans, 171

preparing for metal sash, 183
typical section, *182*
wall fabricating, 173-175
window frame, 174-175
Story pole layout, 171-173, *172*
Stripping steel ply, 72
Strongback, 52, 70
Studs, concrete form, 51
Superform, 83-85, (fig.) 83, 84, 85
Symons slab shore (*see* Flat slab forms)

T

Three, four, five rule, 32-33
Ties, function, 52
Transit, 21-32
    leveling, 25-26, *25*
    nomenclature, 21-23, *22*
    points on a line, 30-32, *30*
    setup, 24-26, *24*
    vernier scale, 29-30, *30*
    wiggle in, 30-32, *31*

U

Uni-Form panels, 72-76, *72*
    assembling, 73-74, *74*
    corners, *73*
    erection, 73-76
    liners, 74-75, *75*
    stripping, 76

W

Waffle slabs, 150-153
    form placement, 151-152, *151*
    forming around columns, 152, *152*
    pan removal, 153, *153*
    stripping, 153
Wales, 52
Wall footing forms, 14-18
    plain, 14-17
    stepped, 17, *18*
    typical, *16*
Wall forms, 35-85
    corner detail, *69*
    erecting, 54-55
    fabricating, 52-55
    panels, manufactured, 66-81
        gates, 67
        steel ply, 67-72
        steel ply erection, 69-72
        Symons hardware, *68*
        Symons panels, *68*
    typical, *51*
    waler installation, 70, *71*
    wedge bolts, 70, *71*
    wire, 39-40
Wallboard, 203-209
Water level, 213-214, *213*, *214*
Window bucks, 188-189, *188*
Wood siding, 175-178
    bevel, 176, *176*
    board and batten, 177-178, *177*
    channel, 178, *178*
    nails, 176, *175*
    plywood, 178-179, *179*
    tongue and groove, 176-177, *176*
Workbench, installation, 247-248, *248*